The Naked Don't Fear the Water

The Naked Don't Fear the Water

An Underground Journey
with Afghan Refugees

Matthieu Aikins

HARPER

An Imprint of HarperCollins*Publishers*

HarperCollins books may be purchased for educational, business, or sales promotional use. For information, please email the Special Markets Department at SPsales@harpercollins.com.

FIRST EDITION

Calligraphy by Roohullah Adib

Library of Congress Cataloging-in-Publication Data has been applied for.

ISBN 978-0-06-305858-3

22 23 24 25 26   LSC   10 9 8 7 6 5 4 3 2 1

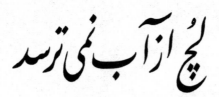

The naked don't fear the water.

—Dari proverb

Contents

Part IV | *The City*

Part I

The War

1

A t first light, I leaned against the window and looked down at the mountains. We were flying into the rising sun, and its rays threw the badlands into relief: corrugated brown cut by green valleys, and speckled with hamlets still reached by donkey. We were near the intersection of Afghanistan, Iran, and Turkmenistan, but which country I saw below I couldn't say. Frost had crystallized on my porthole, rosy with the dawn just like our contrails would be to the people below.

I settled back against the headrest. We were still a few hours out from Kabul, where my friend Omar was waiting for me. When I closed my eyes, I could see his face when he dropped me off that summer at the airport, suddenly pleading, his hand gripping mine: "Come back, brother. Don't leave me. Everyone else is leaving."

The plane was quiet. The few passengers I could see were slumped forward or sprawled out asleep across the rows. These empty places would be filled on the return to Istanbul, I knew, with Afghans fleeing the war. My own seat might be taken by someone who planned to cross the water in the little rubber boats that departed from Turkey to Europe. Thousands of refugees were landing each day now on the Greek islands, and many more were on the way. It was late October 2015, and something miraculous was happening that fall, a violation of a fundamental law: under the weight of the people, the border had opened.

For years, the pressure outside Europe had been building as war spread through the Middle East and made millions homeless. The boat people were mostly Syrian, Afghan, and Iraqi. Many were women and children, and, short of shooting them, there was no way to stop them. From Greece, they headed north through the Balkans, filling city squares and border crossings, a spectacle on the news, a crisis. To keep the European Union from tearing itself apart, Germany suspended its rules and let the migrants through; other countries followed suit, and now the five frontiers between Athens and Berlin were down. Screens around the world showed the masses walking through open borders, proof of the impossible, a clarion announcing universal freedom of movement—a dream for some, and a nightmare for others.

No one knew how long the miracle would last. Thousands of people were landing each day now in the little boats. A million would pass into Europe.

And Omar and I were going to cross with them.

WE HAD MADE OUR DECISION back in August, when I'd returned home to Kabul after an assignment in Yemen. I'd known Omar since I'd started working in Afghanistan, and he'd always dreamed of living in the West, but his aspiration had grown urgent as the civil war intensified and his city was torn apart by bombings. American soldiers were on their way out of the country; I was trying to move on, too, burned out after seven years reporting here, but I couldn't leave Omar behind. So when I'd flown back earlier that summer, my friend had been on my mind. I had no plan yet, but an idea was taking shape. Omar and I needed to talk.

WELCOME TO HAMID KARZAI INTERNATIONAL AIRPORT. At the immigration counter, I handed over my passport and placed my fingertips on the green glow of the scanner, then walked to the baggage carousel and got my suitcase, wheeled it to the X-ray machine. The cop at the monitor was looking for guns and bottles. Alcohol was illegal in the Islamic Republic of Afghanistan, except at the embassies and international agencies, but foreign visitors were allowed to bring in two precious bottles each. I hefted my suitcase onto the conveyor belt, along with the bag of scotch and gin from the duty-free in Istanbul, and walked to the other end, rehearsing my lines.

My ancestors came from Japan and Europe, but I look uncannily Afghan: almond eyes, black hair, wiry beard. So the border guards invariably assumed that I was a local with *haram* contraband, a lucrative catch, since the confiscated booze would likely end up on the black market. Over the years, my Persian got better, but that just made the conversations more awkward.

"Brother, are you telling me you're not Afghan?"

"No, sir," I'd say, scrambling around the belt with my passport before the cop could snatch the bottles. "Look at my name, I'm not even Muslim—sorry."

Outside the terminal, I inhaled the dry summer air. I hadn't been sleeping much since Sanaa, but my tiredness left me as the scene came into focus: faraway snowcaps of the Hindu Kush, the slums on the hillside, the Humvee with its turret pointed at the gate. In the parking lot I spotted a gold Toyota Corolla and, listening to the radio with the window down and a cigarette lit, my friend Omar. He got out and walked forward: taller than me, broad-shouldered, with a fleshy grin and crow's-feet. As we embraced, the heat made his stubble prick against my cheek; he smelled of cologne and smoke. Prying my suitcase from my hand, Omar hefted it into the trunk. We drove into the roundabout outside the airport, a gyre of taxis, armored SUVs, buses, the policemen shouting, the beggars tapping windows, the peddlers swinging racks of phone cards and dashboard ornaments. Omar nosed the Corolla forward, cursing softly, one hand on the wheel and the other clutching a Pine, from time to time leaving it between his lips to run his fingers through his dark mop of hair. It wasn't until we got out onto the airport highway, with its long stretch of cavernous wedding halls, that we could relax and catch up.

"It's good that you're back, *baradar,*" he said in Persian. He smiled but kept his eyes on the road.

"It's good to see you too, brother," I said.

He knew my lease was expiring, and that I'd come back to clear out the house. It seemed like half the city was escaping that summer of *raftan, raftan*—going, going. Afghans were losing hope in their country's future. The middle class spent their savings on flights and visas to Turkey; young men filled buses departing for the southern desert near Iran. Omar's own family was leaving. Four of his siblings were already in Europe, and his mother and sister were getting ready to escape with smugglers. But for a long time, Omar's plan had been to emigrate to America through the Special Immigrant Visa, a program created by Congress to reward loyal Afghan and Iraqi employees—a happy ending for a few, to soothe America's conscience. Omar should have qualified; he'd served in combat as an interpreter for the Special Forces, and worked with USAID and demining contractors. But when he sent his application to me, I saw that he was in trouble. He needed all sorts of paperwork that he'd never thought to collect over the years: letters of recommendation from his supervisors, copies of his employers' contracts with the US government. How was he supposed to track down a Green Beret captain he knew only by first name? Or get documents from the demining company, which had gone out of business? *Hello my dear and sweet brother,* he emailed while I was abroad. *I hope you are fine and doing well. Please wish for me best of luck and find the chance to get the US visa and move there. I am really tired of life here.*

We sent in everything he had. It took two years for the answer to come back: *We regret to advise you that your application for Chief of Mission (COM) approval to submit a petition for the SQ–Special Immigrant Visa (SIV) program has been denied for the following reason(s): Lack sufficient documents to make a determination. . . .*

When his dream of America was dashed, Omar was left with the same prospect as his mother and sister: taking the smuggler's road to Europe, a long and dangerous journey across the mountains and sea. That's when I had my idea. If Omar was going to travel that way, then I wanted to go with him and write about it. Given the risk of being kidnapped or arrested, I'd have to disguise myself as a fellow Afghan migrant, but after all the dangerous assignments we'd done here together, I trusted Omar with my life. This way, I could see the refugee underground from the inside. And I wouldn't have to leave my friend behind. We'd be helping each other. And I would pay for everything.

Omar was silent a moment after I laid it out for him, as we sat parked outside my house that August. He could tell I was serious. Then he grinned. "Of course we can go together."

"Are you sure?"

"I'm sure, brother."

"All right," I said. "When can we leave?"

He sighed. "Not yet," he said. I'd assumed he was ready, but it wasn't so simple. First, he had to get his parents out of the country.

"Of course," I told him.

And there was someone else keeping him here in Kabul: Laila. She was his landlord's daughter and lived two houses down. They'd been seeing each other in secret for several years now, but I hadn't realized things had gotten serious. She was the love of his life, he told me. They planned to get married. But she came from a wealthy Shia family; Omar was Sunni and had only the Corolla to his name. If only he'd gotten the visa to America, he would have had something to offer her family. He could have taken her there legally. Now he had to get asylum in Europe first and then come back for her. But while he was gone, her father might try to marry her off to someone else; Laila told him that she could delay, but not defy, the patriarch's decision.

That was his dilemma: to win Laila, Omar would have to leave, and risk losing her.

AFTER I MADE MY PROPOSAL that day in August, we dropped my luggage off at the house and ran errands. It was late by the time we came back and there was a blackout in the neighborhood, as usual. We had a generator but as we drove up, I could see that the upstairs windows were dark above the courtyard wall, and I wondered if anyone was home; but then Omar honked, and old Turabaz, our *chowkidar*, creaked open the gate for us. As we pulled in, the dog barked and threw herself against her chain.

I'd lived in a few different houses in Kabul during the years I spent there as a freelance journalist, but this was the first I'd made my own. A few years earlier, I moved with three other foreigners. We renovated the house, planted roses in the garden, held parties, and then, one after the other, my friends left the country, replaced by other, increasingly transient housemates. Most expats didn't come to Afghanistan for long. It was an adventure or a chance to make money.

I got out of the car and shone my light on the tufted, yellow lawn. I'd been away for months. The shed, where we'd once distilled vodka, was filled with trash. For security, someone had crudely bricked up one of the doorways that led to the street. And the dog, wild at the best of times, was matted with filth and mad with excitement, her tongue greeting my palms as I crouched to her. "Isn't anyone taking care of her?" I snapped at Turabaz.

Omar was crouched by our old gas generator. We yanked and cursed, but it wouldn't start, so we went from room to room examining the house's furnishings by flashlight. I wanted to sell them and give the money to Turabaz, since he'd soon be out of a job, although the secondhand markets in Kabul were glutted from emigrants liquidating their households. Omar had helped us move in, and he remembered exactly how much we'd overpaid for each item.

"You spent a hundred dollars for that," he said, shining his beam on a dusty pressboard shelf. "It's probably worth five dollars now."

When Omar went to check out the kitchen, I sat down at a desk in the living room. I was starting to feel the jet lag. We used this room as our office, and I'd written a lot of my stories here, with a gas heater hissing in the winter, the door open to the garden in the summer. In the gloom, the carpet's stains were faintly visible. I rubbed one with my toe—red wine. When we hosted parties we pushed the desks together into a bar that grew sticky with homemade punch. People from all over the world had danced together here. For a while we'd called this country home. Now we were leaving it like a shell we'd outgrown.

When we finished our inventory, Omar and I took the dog for a walk. Turabaz had named her Baad, which means "wind" in Persian. She was mostly German shepherd, I think, and I liked to show her off because home invasions were becoming a problem. When I walked her, the kids in the street, seeing her daggerish grin, cried *gorg*, wolf. She was affectionate, but difficult to train due to a tic from some puppyhood trauma. At the slightest pressure on her hindquarters, she'd chase her tail in a snarling loop that brought to mind the self-swallowing serpent Ouroboros. One of my since-departed housemates had acquired her on a whim while I was out of town. I still had to figure out what to do with her.

Kabul's streets were empty at night. We walked over to Kolola Pushta Hill, a pair of mounds with a graveyard on one and a mud-walled fort on the other, built by the British in the nineteenth century and now home to an Afghan army unit. As Baad snuffled at a gutter, Omar stalked ahead, whispering into his phone to Laila. He was telling her what he'd told me as we were driving home. He'd made up his mind to leave and become a refugee, but not until he and Laila were engaged. He was going to ask her father for her hand, on the assumption that Omar could get asylum and bring his bride to Europe. He'd warned me that it might take some time to convince the patriarch. I replied that I could be patient. I had to go back to the US anyway to finish an assignment but I planned to return in October. Surely Omar would be ready by then.

The track wound upward among the gravestones, jagged stones with sticks and rags tied to them. Across from us, the outline of the fort sunk against the streetlights beyond. A scraping cough came from the darkness of the graveyard, and then the smell of hashish. I tightened my grip on Baad's leash. Let Omar try to win his beloved, I thought. If we were going to travel underground together, then I needed time to prepare to pass as Afghan. Once we started there would be no turning back, not without abandoning my friend. Because we might be searched, I'd have to leave behind the American and Canadian passports that allowed me to move so easily through this world full of borders. And yet it wasn't just checkpoints and fences that governed our movements; there were laws and webs of surveillance and more intangible lines drawn by self-interest—the tracks our lives ran on, the limits to our imagination. *The wall is also inside each one of us,* John Berger wrote.

At the top of the hill, there was an empty lot ringed by trees. I walked to the edge and looked north, where I could see clear out past Qasaba, where the slum crept up the steep hills that enclosed the capital. The power had come back; many of the makeshift homes were electrified now. After Omar finished his call, he walked over and stood beside me.

"When we first came here, there were no lights," he said.

Like so many Afghans from his generation, Omar had grown up as a refugee in Iran and Pakistan. In 2002, his family had returned from exile to a shattered capital, driving down the avenues of rubble past buildings whose shell holes were screened by ragged curtains. But the people had hope. Kabul had grown in spurts of concrete, sprouted shopping malls and neon-fronted gas stations, but the promise of peace had been a lie. The war that raged out in the countryside was drawing closer to the capital. The Taliban were coming. And yet at night you couldn't see the blast walls topped with concertina wire, or the unpaved streets where widows begged come morning. The city before us was made of light.

"It's beautiful," I said.

"It is. And, God willing, it will get better one day."

"But you're ready to leave?"

When he turned to me, I could see he was tired.

"There is no future for me here. You have a good job, you have documents, you can travel anywhere you want." He looked out at his city. "The only thing I have is my luck."

2

I left soon afterward for New York and when I returned three months later at the end of October, on that empty plane via Istanbul, I found the baggage carousel in Kabul crowded with men in white robes unloading containers of Zamzama holy water, which they'd brought back from Mecca. The hajj pilgrimage in 2015 had been a disaster, with more than two thousand killed in a stampede, and another hundred by the collapse of a crane belonging to the Saudi Binladin Group.

I got my booze through the scanner and went to find Omar in the parking lot. The guards at the airport seemed on edge; a few weeks earlier, the Taliban had captured Kunduz, a border city near Tajikistan. The government's defenses crumbled under the sudden assault and, for the first time since 2001, the Taliban raised their white banner in a provincial capital. A stream of the displaced headed south to Kabul, spreading panic as they went. The fall of Kunduz added momentum to Afghanistan's exodus, already at a fever pitch since the border had opened in Europe that fall.

As we drove away from the airport, I started to tell Omar about the so-called humanitarian corridor that had been opened for refugees through the Balkans, but he knew all about it from watching the news at home. A miracle had cleared the way for us, and yet he told me that he still hadn't made his proposal to Laila's family, or settled his own parents' departure. It was complicated; he needed more time. But it was OK, I told him, because I wanted to do one last story together in Afghanistan. A shocking incident had taken place during the fall of Kunduz: A team of US Special Forces, battling to retake the city alongside Afghan troops, had called in an airstrike on a Doctors Without Borders hospital, killing forty-two people. The military claimed it was an accident, but I knew that local authorities had long held a grudge against the hospital for treating wounded insurgents. I wanted to investigate, and I needed Omar as a driver. We'd go to Kunduz together, and then I could finish writing the story while he sorted things out with Laila. We didn't have to rush things. I was confident that we were going to leave Afghanistan together, no matter what. Our trip would close a circle, for there had been a reciprocity in our motion, it seemed to me, since the day we met.

I HAD BEEN WORKING WITH Omar since my first magazine story in Afghanistan, more than six and a half years earlier. It was the spring of 2009 and I was twenty-four. I'd just gotten an assignment from *Harper's* to write a profile of Colonel Abdul Raziq, a border police commander who was a key ally of the US military and, it was rumored, in league with drug traffickers. I wanted to go to Raziq's frontline province of Kandahar, but the magazine couldn't afford any of the capital's established fixers, who were charging hundreds of dollars per day to work in the dangerous south, if they were willing to go at all.

I was staying at the Mustafa Hotel in downtown Kabul, and when I explained my predicament to Abdullah, the lugubrious manager, he said he knew the right guy, a former military interpreter who was also getting started in journalism. So one day I walked into the lobby and there was a kid about my age waiting for me: Omar. He jumped to his feet and clasped his raspy palm on mine. "Nice to meet you, bro," he said. "I'll go to Kandahar with you, no problem."

It was midday, and he asked if I was hungry. We went out to the cordoned-off street that the Mustafa shared with the Indian embassy, a location that protected guests from kidnappings but exposed them to the occasional car bomb. Omar's Corolla was parked nearby. It was a short drive to the restaurant but traffic moved at a crawl over the rutted, dusty streets past Shahr-e Nau Park.

"Kandahar is fucked up," he told me. His nearly fluent English was larded with the locker-room expressions he'd learned from the soldiers. "I've been there with the coalition forces." He'd been working in the south for a few years now, on contracts with the Americans, Canadians, and British. He was getting tired of the dangerous patrols and the tedium of life on base, and wanted to work as a freelance fixer in Kabul, which was back then teeming with foreigners.

Like me, Omar's adult life had been coeval with the war on terror. He'd grown up in exile, and he and his family had returned soon after the American invasion, eager to take part in the promised era of peace and reconstruction, but the country was in ruins and jobs were hard to find. He'd heard the foreign troops were paying good salaries to do dangerous work down in Kandahar; finally, in 2006 he took the bus without telling his mother where he was headed.

He didn't speak much Pashto, the language of the south, but there was a shortage of English-speaking locals and he was hired right away by one of the companies that supplied interpreters to the foreigners. Omar's first assignment was with the Canadians; his starting salary was six hundred dollars a month, six times what an ordinary Afghan soldier made. He and the other translators lived on the giant base that had sprouted in the desert by the airport, behind miles of earthen Hesco barriers and concertina wire, in a grid of housing containers and dusty gravel that threw back the harsh sunlight. Omar was dazzled by the hulking armored vehicles and the jets that rattled his teeth as they landed, the generators that guzzled fuel night and day to power air-conditioned tents, and the endless pallets of soft drinks and frozen steaks hauled here by jingle-ornamented trucks from ports in Pakistan.

Omar had observed Westerners on television since he was a little kid but this was the first time he'd gotten up close. He, like the other *terps*, learned to embody their trustworthiness by adopting the soldiers' slang, clean shaves, and shades; their respect for rules; their attitude toward the *bad guys*. It was easy for Omar because he liked the Canadians. He knew they came from a land of plenty, but they seemed far more generous and honest than the people he'd grown up with as a refugee in Iran and Pakistan, where hardship and fear could turn kin against one another. The Canucks shared their stubby cigarettes and gave him winter jackets and boots made with synthetic materials he'd never touched before. *Their eyes were full*, as the Persian expression went. The foreigners said they came to fight terrorism and to help his country. Omar believed them.

But the Taliban were on the rise in the farmlands surrounding the city. From a helicopter, the Panjwai valley looked stark green against the desert, its mud-colored canals shaded by mulberry trees. There were rows of pomegranate orchards, each earth-walled plot with a cow and a few sheep and a guard dog, worked by subsistence farmers, tenants mostly. Sweating under helmets and body armor, the Canadians walked down embankments whose softness might hide jerricans of homemade explosive, *leg lottery*, they called it. The dogs sometimes had to be shot when they raided the little compounds, where they searched a couple of tin trunks and some bedding, probed the courtyard with bayonets and metal detectors, while the women and children sobbed quietly beside sullen youths with suspiciously soft hands, and old men who'd once watched the Soviets with the same hooded look.

The Canadian infantry patrolled in strength by day, accompanied by the Afghan army and police, but night belonged to the insurgents and to the foreigners who hunted them, the bearded men Omar saw sometimes with blindfolded captives, one of those things he knew never to ask about. The Taliban took prisoners too, who were judged by mobile sharia courts; collaborators like Omar were marked for assassination. Three of his fellow interpreters were shot outside the city, another five killed when their bus was hit by a bomb on the way to base. His mother begged him to quit, but he needed the money and kept going back to Kandahar and Helmand, working stints with the Royal Marines and Green Berets. The terps weren't given combat training but they were part of the war. Not long after he started, he experienced his first battle when the Canadians launched an offensive into the valley west of Kandahar City. His platoon was sent to hold some earthen berms in the middle of grape orchards. On his second night spent trying to stay warm inside an armored vehicle, a soldier told Omar to get out, and handed him a rifle.

"Can you defend yourself with this?" the Canadian asked. He sounded worried. "There are a lot of bad guys around."

Omar gripped the cold plastic of the C7. In Iran, he and his classmates had learned to shoot Kalashnikovs, in case the Americans ever invaded. This rifle wasn't so different.

They put him in the perimeter with the rest of the platoon, thirty-odd men and a female medic. Out there in the night were an unknown number of Taliban, massing to overrun their isolated strongpoint. He crouched behind a berm. Someone yelled that the insurgents were trying to flank them, and the gunfire started: the incoming cracking overhead and the Canadians' deafening return fire, the 25-millimeters on the vehicles booming like pile drivers.

Omar fired his clip into the darkness. His ears rang and he tasted gunpowder. Finally, they heard the long arriving roar of the jets. A bomb strike lit up the night, showing the faces around him. At dawn, the Canadian tanks arrived, the earth rumbling as they passed. When the battle was over, the platoon moved forward and found the bodies in the vineyards and shattered farmhouses, youth in blood-soaked robes and ammo bandoliers. His countrymen.

IT'S HARD FOR ME TO recall Omar as the stranger he was that spring day we first met in 2009, telling me about his time in Kandahar over lunch. I do remember how lush the restaurant's garden was, where we sat and ate skewers of grilled mutton. When Omar asked whether it was my first visit to Afghanistan, I explained that I'd visited the previous fall, during a backpacking trip through central Asia.

After graduating from college in 2006, I'd moved home with my parents in Nova Scotia. I wanted to be a writer and I thought I'd find in the world the material I lacked within myself. After a couple of years of working odd jobs, I'd saved enough for a one-way ticket to Paris in the spring of 2008. I hitchhiked to the Balkans, copying road maps into my notebook, alongside the names of cities that I spelled out in block letters so I could hold the pages open on the side of the road for drivers to see: TRENTO, LJUBLJANA, NOVI SAD. I spent the summer in Croatia, swimming muddy rivers and drinking plum brandy with a group of punks who'd picked me up at a music festival. I slept on couches and said yes to whatever came my way. When autumn arrived I decided to travel overland to India and oriented myself through central Asia, which meant I had to pass through either Turkmenistan or Afghanistan. In Tashkent, it turned out to be easier to get an Afghan visa, so in October I walked south across the Friendship Bridge, where the last Soviet tanks had retreated two decades earlier.

The Amu Darya ran wide and silty below. I hadn't made it halfway across when a driver slowed for me, a trader on his way back to Mazar-e-Sharif, the city where I was headed. Most people in northern Afghanistan spoke Dari, a dialect of Persian, and I tried out the first dozen words in the phrase book I'd picked up.

The road ran south through an expanse of gray dunes, where tent camps and herds of camels faded in and out of the distant haze. When we reached the villages on the outskirts of Mazar, I stared out the window at the mud-walled houses and bearded men in turbans. In the concrete ex-Soviet cities I'd just left, people had been drinking vodka in the cafés, even during Ramadan. I was most surprised to see women in head-to-toe burqas: Hadn't that garment been defeated along with the Taliban?

Mazar was centered around the Blue Mosque, whose gates and grand domes were tessellated with thousands of turquoise tiles. Local legend had it that Ali, son-in-law of the Prophet Muhammad, was buried there. I found a hotel on the south side of the square called the Aamo. It was a three-story wreck frequented by truckers and pilgrims, the halls littered with tea dregs and cigarette butts. For ten dollars I got a room to myself with four worn beds that overlooked the Blue Mosque. That night, I sat at the window and tried to survey it all: although the square was lit up in neon like a Vegas casino, complete with blinking palm trees, it was deserted, and I couldn't shake a sense of melancholy, thinking of the raw poverty I'd seen for the first time that day, little kids ragpicking amid the sewage.

My arrival was a source of great entertainment for the group of young men who worked at the hotel. There was Jawed, who put his hennaed hand in mine as we walked in our socks across the marble courtyard of the mosque. Kamran, the buff one, took me for ice cream and french fries, and tweaked my wrist after he insisted we arm-wrestle. Ibrahim, with hazel eyes and a push-broom mustache, spoke the best English and ran the front desk.

"Do you know Brian Tracy? I think he is very famous in your country," he said. Ibrahim was reading a self-help book called *Eat That Frog!* "It shows you how not to waste your time." In his spare hours, he studied business management and computer programs. He asked me how he could emigrate to Canada. I had no idea. Was it even possible? Ibrahim knew one way: he was saving money to hire a smuggler. *It's really simple*, I wrote in my diary. *I come from a place they would like to go, but can't. It would drive me crazy but for them it's just matter-of-fact.*

Even working in restaurants or construction sites back home, I had earned more in a day or two than they did in a month. There was a gulf between us, but I thought we could bridge it in our encounter as humans. In spite of all our cultural differences, I felt at ease hanging out with them. The intimacy they offered was different from the kind of male bonding I'd grown up with, the drunken roughhousing and predatory banter—here women were barely mentioned and they were not seen at all in public places like the *chaikhana*, the teahouses on the square. Afghan men were openly affectionate with one another. It was as if, having excluded women, they'd apportioned a surplus sociality among themselves. The boys took me to the bazaar and helped me buy a ready-made *piran tombon*, the knee-length tunic and baggy pantaloons that are the traditional garb in Afghanistan. When I unwrapped the pants in my room, I burst out laughing. I thought we'd bought the wrong size; the waist was as wide as my arm span. But no, Jawed came and showed me: you just cinched them tight with a cotton band, and let all that fabric balloon around you. When I came down dressed up, the boys hooted in approval. They wound a black-and-white-patterned scarf into a turban around my brow and then stared in amazement at how Afghan I looked.

With my dark hair and Asian eyes, I had crossed a color line somewhere over the Atlantic. In Europe, I was no longer included with the whites. I got called a *Paki* in England; in France, I was *arabe*. But as I traveled into central Asia, it was like walking toward a mirror; in northern Afghanistan, with its mix of Hazaras, Tajiks, and Uzbeks, I'd found my phenotype. People saw their face in mine.

At the teahouses, the boys amused themselves by calling over a passing friend and, motioning for me to be silent, making him guess which province I was from. "He's a foreigner!" they exclaimed at last. "But why does he look so Afghan?" the other asked in amazement.

In halting Dari, I explained that my father is European Canadian, and that my mother was born in America, but her grandparents were Asian. "Japan is here, Canada is there," I said, holding two fingers apart, then bringing them together and grinning. "And Afghanistan is in between."

From Mazar, I took the bus to Kabul, where I checked into the Mustafa, a blocky high-rise built in the 1960s that had been a popular stop on the Hippie Trail. During Taliban times the Mustafa was converted into an indoor bazaar but, after the US invasion, it was one of the first hotels to reopen, and anyone who couldn't afford the Intercontinental stayed there. By the time I arrived in late 2008, there were plenty of cleaner options, but at twenty bucks a night, the Mustafa was the cheapest one safe for foreigners. The clientele came from the bottom rung of expats, unemployable contractors, freelance humanitarians, and wannabe writers like me. Sitting at the pink onyx bar, I met a leathery soldier of fortune who told me he was from Rhodesia. Another guest, a sad-eyed Swiss correspondent, said he'd spent the past decade battling a heroin addiction. He and I were sitting in the lounge, listening to the street dogs howl after everyone had gone to bed, when he brought out his pipe and showed me how to melt a ball of opium and inhale the fragrant vapor, which spread through my limbs and floated me to bed through hallways of mirrored bazaar stalls.

My monthlong tourist visa was running out, and I wanted to go to Iran next, but the main road, Highway 1, which ran south through Kandahar to the border, was too dangerous—the Taliban might stop the bus and kidnap you. But flying was for tourists. There was a less-used route that crossed the spine of the Hindu Kush; however, you had to take local transport and stay in teahouses which doubled as roadside inns. Since Mazar, I'd told people I was Kazakh whenever I felt unsafe, and I decided to pretend I was a migrant laborer headed to Iran in search of work.

I was scared, but once I climbed into the van in Bamiyan there was no turning back from the lie. For days, we followed dirt roads up through the snow line. This was the roof of the world, mountain chains that stretched all the way to Tibet, and the panorama seemed like a dream I couldn't interpret. I kept calling attention to myself by making blunders, like peeing standing up instead of squatting, or praying like a Shia when I'd said I was Sunni. Still, as strange as I must have seemed as a Kazakh migrant, none of my fellow travelers guessed the weirder truth. People were suspicious of one another anyway, and hid their origins and destinations, afraid of what could happen on a road menaced by bandits and insurgents. At night, when we lay down in rows on a teahouse floor, the travelers whispered about recent be-headings and abductions. Like a lot of kids who'd grown up safe, I'd been curious about death, and here it was, all around us.

A week after I set out from Kabul, our van rattled down a river valley and into the border city of Herat. Relieved that I had made it there alive, I splurged on a hotel with hot water. As I stood in the shower, I kept the door open so that I could watch the TV in the corner of the room. Barack Obama had just been elected president of the United States. He was giving his victory speech in Chicago. I turned the volume up so that his voice rose above the hiss:

And to all those watching tonight from beyond our shores, from parliaments and palaces, to those who are huddled around radios in the forgotten corners of our world, our stories are singular, but our destiny is shared, and a new dawn of American leadership is at hand.

I closed my eyes under the water.

AFTER OMAR AND I FINISHED introducing ourselves over lunch the following spring, I came to the business at hand. "Do you know who Abdul Raziq is?" I asked. Of course he did. Raziq wasn't famous across the whole country yet, but anyone who'd worked in Kandahar knew about the boyish, ruthless, thirty-year-old master of Spin Boldak, the main border crossing with Pakistan in the south. Already a colonel in the border police, Raziq was feared and admired for his take-no-prisoners attitude toward the Taliban, but there were persistent allegations that he was trafficking opium and murdering his tribal opponents. Investigating him could be risky. And there was something else I had to tell Omar: I'd already been to Spin Boldak with Raziq's men, on a quasi-undercover journey.

The previous fall, after I had made it across the country to Herat, I entered Iran with my Canadian passport and spent a couple of months traveling there. I returned to the pleasures of backpacking—a visit to a nearly empty Persepolis, a Christmas swim in the Strait of Hormuz—and Afghanistan's war faded from my mind. I started thinking about applying to grad school, for journalism maybe, and still thought the end of my journey lay in India. To get there overland I had to cross Pakistan via the border city of Quetta, a dangerous place—an American working for the UN was abducted there around that time. The day I arrived from Iran, I was walking down the street, feeling conspicuous in Western garb, when a luxury SUV pulled up. The tinted window rolled down, and two young men in robes, with shaggy haircuts, beckoned me over.

"Oh, you're a tourist?" one said in English, smiling. He invited me to join them for lunch. I hesitated, but something about their frank curiosity suggested they weren't going to kidnap me. I got in, and we spent the next week hanging out together, smoking hash and shooting guns. They began to tell me secrets, like the mistresses they kept hidden from their families. Knowledgeable about Quetta's underworld, my two friends showed me the hospital where wounded Taliban were being treated. Pakistan was supposed to be a US ally, but the military was playing both sides by supporting the insurgency in Afghanistan. Quetta seethed with shadow wars: sectarians killing Shias, Baloch separatists attacking the government, mafia and tribal vendettas.

My hosts were scions of local Pashtun clans that were deeply enmeshed in smuggling; the British had drawn a border through their extended families about a century earlier. Some of their kin were in the Taliban; one uncle was the deputy chief of police in Quetta. My new friends were proud of their success, and explained they were shipping two metric tons of opium from Afghanistan to Iran each month, making around a quarter-million dollars in profit each time. A heavily armed convoy of Land Cruisers sped through the stony wasteland that was the nexus of three countries. The Iranian border guards were dangerous and had to be evaded, whereas the Pakistanis were easily bribed, they said. But the Afghan connection was the most important one.

"Who is it?" I asked.

"The big boss. Abdul Raziq."

I had told them I was a writer, but not an aspiring journalist. When they explained that Raziq was supported by the American military, I smelled a story. My hosts had bought a baby tiger as a gift for Raziq, and I asked them if they would take me with them, the next time they went to Spin Boldak. They agreed; I'm not sure why, except out of friendship and boredom. They had trusted me, as I trusted them. They knew the cops at the border, and it was easy to cross into Afghanistan without having to show my passport. We passed through poppy fields on the outskirts of town; the country supplied nearly all of the world's illicit opium. The border generated profit; Spin Boldak had a shantytown of shipping containers converted into shops and dwellings, where men and boys toiled to unload secondhand microwave ovens, guitars, DVD players, bicycles, propane stoves, motorized wheelchairs, generators, children's toys, and cars, lots of used cars. Most of these goods were imported cheap into Afghanistan and then smuggled back to Pakistan, where duties were high. The smugglers and the police were often the same people and Raziq's men were taxing everything, legal and illegal.

I had to spend ten days waiting there until Raziq returned for his maternal grandmother's funeral. At the ceremony, my friend pointed out one of the guests, a burly, bearded man: "That's Rahmatullah Sangaryar," he whispered. "He was in Guantanamo."

I approached the dais; Raziq looked even younger than his thirty years. He had a close-cropped beard and a widow's peak that poked out from his cap, and was dressed in a simple white robe and pinstriped waistcoat. He shook my hand, smiled politely, and turned to the next visitor. I went back to Pakistan.

I had proof of Raziq's connection with drug traffickers, but I had to do more reporting. I got an Afghan visa in Pakistan, flew to Kabul, and checked into the Mustafa, where I first met Omar.

After I explained the whole story, I said I'd understand if Omar changed his mind about going to Kandahar. Raziq might not be happy to see me again. But Omar didn't flinch. We flew south together, where he did his best to translate Pashto, and I did my best to translate our interviews into a magazine article. Sometimes, hearing us speak together in English, the locals guessed that Omar, in his T-shirt and wraparound shades, was the foreigner, and that I in my robe was his interpreter, something that amused him no end.

We heard lots about opium trafficking but also darker stories of bodies dumped in the desert by Raziq's men, bearing signs of torture. In explaining Raziq's sudden rise, people kept coming back to his close relationship with the Americans, and his frequent visits to the CIA and Special Forces' base in town. America needed allies, particularly on strategic routes like Spin Boldak. The first combat brigades sent by the new president had already deployed in the south; by the end of the following year, the number of US troops in Afghanistan would triple. "This is a war we have to win," Obama said.

At night in our hotel in Kandahar City, Omar and I lay in our twin beds in the dark and listened to the gun battles raging in the suburbs. The Taliban were on the city's doorstep. He and I tried to untangle the stories we heard that day, about tribes and blood feuds and business deals, which explained the fractal pattern of the war better than the binaries I'd arrived with, police and criminals, the Taliban and the government, and the West and terrorism. But how to explain it to the people back home who were interested, belatedly, in this faraway country they'd invaded? When we grew tired of discussing the day's work, Omar and I talked instead about ourselves, and of the past and future, to the drumming of distant guns.

"What's Canada like, bro?"

"It's cold."

"That's OK," he said, and I could picture his eyes gazing up into the dark. "I like the cold."

3

On our final reporting trip together that fall of 2015 to investigate the American airstrike on a hospital, Omar seemed distracted. He was up all night on his phone. He backed the car into a wall outside the provincial council in Kunduz. When I left him with Victor, our photographer, they got briefly detained after they bumbled into a commando operation. And Omar kept playing his favorite sad song, "My Heart Will Go On," on the car stereo until finally Vic and I made him stop. Then he put on headphones.

Kunduz was a combat zone and we had to stay focused, but once we were on our way back to Kabul, I asked Omar what was bothering him. He said his mother had gone to present his suit to Laila's father, their landlord, who heard her out politely but rejected Omar, saying that his daughter was still too young. "She's already nineteen," Omar seethed to me. "It's just an excuse. There will be many suitors for her."

I asked Omar what this meant for our trip to Europe, but he said he didn't know. He needed more time; he had to try to persuade her father. "I can't leave without getting engaged," he told me.

In Kabul, Omar dropped me off at the house where I was staying, now that I'd moved out of my place. I poured myself a drink and went out into the rose garden. Poor Omar. He was Sunni and, worse, he was broke. He wanted to marry a rich Shia's daughter, but there was no logic to love. Yet I'd never seen him like this. He'd had a string of girlfriends over the years, not easy in a place as conservative as Kabul. He'd dedicated himself to chasing good times, back in the days when the capital was flush with easy money. But something had changed with Laila. I didn't think much of it when I first heard about her, after I had moved into my old house. It must have been the fall of 2012, around the time when we had a big party, although my housemate Bette and I had called it a *reception*, in order to distinguish it from the typical Kabul booze-up.

During the surge, the expat bubble was filled with embassy soirees, barbecues at UN agencies, welcome and goodbye parties with themes like *Tarts and Taliban*. Our event was going to be a classy affair. That afternoon, Omar and Turabaz had hung up strings of colored lights and stacked wood in our firepit. We'd stashed the alcohol in the kitchen, out of respect for the Afghan dignitaries we'd invited, but there were soft drinks in tubs of ice under a long bench table laden with fresh vegetables from the bazaar, along with the bloodred pomegranates that were in season.

Bette was a Dutch freelancer who traveled the south in a burqa, interviewing the Taliban. She and I had social ambitions that night. We wanted to show we could draw as good a crowd as the newspaper bureaus; we'd invited generals and ministers, Afghan celebrities and foreign diplomats. But would anyone show up? Various advance security teams for the VIPs had looked over our house, grimly noting the lack of a safe room or armed guards—all we had was the dog and a shotgun I kept under my bed. Kabul was wracked by suicide bombings and kidnappings, and many of the internationals weren't allowed off their compounds anymore. But if enough bigwigs came, the party would have its own guard force.

That spring, I'd moved in with three friends: Bette; Elsbeth, who was also Dutch and worked for an NGO; and Misha, a Russian photographer. We'd found a two-story house with a little courtyard in Qala-e Fatullah, close enough to the Green Zone that Black Hawks would roar overhead on their landing approach. Rent was cheap but the place needed work, and Misha and I squatted for a couple of weeks on bare carpets while Omar helped us wrangle the carpenter and plumber, our hair shock-stiff from the dust the mason kicked up cutting the new countertop, black-and-white-flecked marble from Herat.

Our party was on November 14, 2012. Obama had just been re-elected. We were in the third year of his surge which, at its peak, had a hundred thousand US troops in country, plus an equal number of contractors doing everything from security to plumbing, and on top of that, another forty thousand allied forces—a force double the size of the Soviets', led by America's best and brightest, warrior scholars who'd read *Three Cups of Tea*.

Employ money as a weapon system, General David Petraeus had counseled. By that point, the US had spent half a trillion dollars on the war. An entire economy was created around foreign money, with locals at the bottom of the pyramid, digging ditches and driving cargo trucks. Then you had what the military dubbed *third country nationals*, which usually meant people recruited from poor and middle-income countries, like Filipina accountants and Nepali guards. At the pinnacle were the expats who, by virtue of their English, Western degrees, and personal connections, commanded *international* salaries at the big NGOs, contracting firms, and UN agencies. They were the ones inside the armored SUVs that shuttled around town, mostly male, mostly white; some trailing conflicts and disasters for decades, others fresh out of college, enjoying tax-free salaries and the sudden jump in seniority that a war-zone job offered. Back home in the States, people were still reeling from the Great Recession, with twelve million unemployed, but here you might find yourself with a six-figure job, with a free house, a chauffeur, a cook, a gardener, a gatekeeper, and a maid.

As freelancers, we weren't quite living that lifestyle. When our guests started showing up, I told Turabaz, who couldn't read, to pretend he was checking their names off a list. Darkness had improved the rustic charms of our courtyard, lit now with the tiki torches and Christmas lights. I asked Omar to start the bonfire. The local musicians, seated cross-legged on the carpets, played some classical Persian melodies. Bette was holding court near the kebab table, where the *ustad* laid skewers on a charcoal brazier that he stoked with a reed fan. Omar sidled up to me and murmured, "The musicians want some whiskey in regular cups, not glasses."

I glanced over at the *rubab* player, a mustachioed gallant in an embroidered vest, who winked. "OK, follow me."

Pushing my way through the crowd by the fridge, I mixed some Ballantine's with Coke, handed the coffee mugs to Omar, and then went to check on Turabaz at the gate.

"How is it out there?"

"See for yourself."

I peered out. Our street was choked with armored cars and pick-ups. There was nearly an infantry platoon's worth of firepower out there: troopers in tiger stripes, a British army security detail, expat contractors with accessorized bullpups, Kandaharis with glitter tape on their rifle stocks.

I walked back through the courtyard, where the flames cast a mob along the wall. I spotted Dr. Abdullah, the perpetual runner-up presidential candidate, bent over Nancy Hatch Dupree, at eighty-five the grande dame of Afghanistan scholars. Our reception was a success.

When the dignitaries—at least the ones who might have been offended—left, the booze and hash came out, and Baad was unleashed to frolic among the fallen kebab. The party would last until the muezzin called. But was that the night that we took everyone out to the shed to admire the pot still, branded "Katyusha," that Misha brought back from Moscow? Or the night the head of a UN agency lost his wallet, with his security badge inside, while grinding to "Call Me Maybe"? Either way, we crowded into the foyer to dance, where a disco light spun, and raised the volume and our glasses to drown what was beyond the courtyard wall, the war that was getting worse, our collective failure, and the fact that this wasn't our home and that maybe we had none, at least not together.

And where was Omar? After the musicians and dignitaries went home, I plied him with whiskey and backslapping, told him to dance, to enjoy himself, to pick someone up. Our other Afghan friends were doing it, but he came from a different social class. From amid the press of bodies in the foyer, I spotted Omar leaning against the wall, his glass in hand, a faint smile on his face.

OMAR LIVED TEN MINUTES AWAY from me, in a house that he and his family rented. Their yard was full of overgrown fig trees, but even through the summer foliage he could make out the roof of the three-story dwelling where his landlord, a prosperous businessman, lived with his wife and two children. Shortly after Omar moved in—this was around the time of our first meeting at the Mustafa in 2009—Omar bumped into his landlord's daughter in the alley, a teenager with pale, fine features like a miniature on porcelain. Laila was still young enough at the time to speak to him in public without fear of scandal, and she asked if he was comfortable in his new home. Later, when he and his mother paid a courtesy visit to their neighbor, Laila came in with a tray of tea and candied nuts, and Omar felt her eyes on him.

He was too old for her so it amused him, nothing more, how she watched him from then onward, spying from the roof of her house when he did bicep curls in his yard, or peeking with her girlfriend through a side door which shut as he passed, their peals of laughter ringing through the metal.

Confined to home and school by her strict father, Laila saw Omar come and go behind the wheel of his Corolla, a 1996 gold-colored, four-speed automatic shipped off to Afghanistan after a decade of driving Canadian roads. We drove that car all over the country together, slaloming past trucks on the switchbacks down the Mahipar, around bomb craters on Highway 1, and over the dirt tracks that ringed the sky-blue lakes of Band-e Amir. Omar was mad to drive in those days; *maraz-e motorwani*, he called it: driving sickness. He felt free when he was on the move, a world of possibility before him. In part, his freedom was economic; even if his gigs with foreigners dried up, which happened from time to time, he could always moonlight as a taxi driver.

His city, Kabul *jan*, dear Kabul, was where he most loved to roam. Maps were a language he'd never been taught but he knew how the streets fit together, which ones flooded during rainstorms and which seized with traffic when there was a bombing downtown. He knew shortcuts through the graveyards behind Carte Parwan and along the garbage-strewn riverbank at Pul-e Surkh. Omar drove with one hand on the peeling vinyl of the wheel and held a Pine with the other, his chin swaying to the cassette deck, his soundtrack a mix of old and new, Enrique Iglesias hits like "Hero," and classics by Ahmad Zahir, the Elvis of Afghanistan, who'd crooned for Omar's parents although the lyrics were often from much older poets, the great mystics like Hafez of Shiraz:

> This heart came into life without you,
> It is time for your return.

The Sufi poets spoke of their longing for reunion. Life itself was a form of exile, an estrangement from divine love. They wandered the earth in search of the beloved.

> The pain of separation soothes me in this lonely bed,
> The memory of union, my companion in this empty corner.

The mystic's beloved was God, but one whose truth permeated existence, transcendent and yet immanent within the world and ourselves. Human beauty could reflect divine beauty; our love for one another could kindle our love for God. The Sufis were Muslims, of course, but this idea of love exists across many faiths; the Jewish scholar Martin Buber once wrote that *the You of her eyes allows him to gaze into a ray of the eternal You.*

In a society fenced by tradition and gender, Omar believed that love was freedom. And yet he felt enslaved by desire. When he was ten years old and a refugee in Iran, an older girl from his neighborhood bribed him so that she could play games with their bodies; that was his first time. Now he burned with an unquenchable thirst; his movements through the city were impelled by longing. When he saw a pretty woman hailing a ride, he'd stop and negotiate a fare. They could

chat the rest of the trip, if she wanted. He was polite and unthreaten-
ing, quick with jokes, and attractive—he looked a bit like a young
Ahmad Zahir, in fact, with his wavy locks and full jaw. If they got
along, he'd offer his number, or take hers. Sometimes the girls called
him; sometimes the numbers they gave him were real. A small fraction
of a large number of rides. They called at odd hours, whenever they
could get away from their parents or teachers, and I could tell when
they did because Omar's voice would get all hushed and dreamy.

He built relationships by the minute, out of piles of tattered phone
cards. At last, the girl would agree to meet—but where could they go,
when sex before marriage was not only taboo but illegal? They could
sit chastely at an expensive coffee shop, but there was nowhere to be
alone together. He couldn't bring her to his family home. At a hotel,
the clerk would ask for the *nikah nama*, the marriage certificate. A
father or brother who caught him inside their own home would be, in
the eyes of the public, justified in killing him. The cops were the biggest
danger—if they caught an unmarried couple, they'd extort a bribe, or
even try to rape the girl.

And yet, driven by a force stronger than fear, Omar found places:
a walled orchard in his mother's village or, on a Friday, an empty office
where his friend had a key. At their most desperate, they'd cruise the
city in the Corolla, touching in quick swoops. Once you knew where to
look, Kabul was full of such couples, out of place like migratory birds,
the girl beneath a headscarf, the boy with his eyes on the rearview
mirror.

All across the city, I don't see a single sober face, crooned Zahir, re-
citing a poem by Rumi. *One's worse than the other, everyone wild and
crazed.*

Some Sufis believed that by debasing themselves in the eyes of the
world, they could leave behind their false piety. Wine washed away
the stains of hypocrisy; naked of their ego, they could know God.

> *Come, my dear, to the tavern in the ruins, to see the delights of
> love.*
> *What joy is there, my dear, besides conversing with the beloved?*

There were some women whom Omar treated as disposable, the ones he recognized waiting on the side of the road, who responded to the flicker of his headlights. But he repented afterward. He believed true love would lead him to a better life and so he didn't want an arranged marriage with a stranger. He wanted modern love, as enacted onscreen by heroes like Leonardo DiCaprio or Aamir Khan, a love both compelled and chosen. He drove through the city, waiting for her face to emerge from the crowd, while Zahir sang Rumi's words:

All you who've gone to hajj, where are you? Where are you?
The beloved is right here, come, come.

In our exile, the face we seek is veiled from us. At which door should we stand, waiting for love like revelation, listening for the jubilee call?

Your beloved shares the same wall with you,
What are you hoping for, wandering the desert?

But the few times he fell in love, the girl had been the one to end things, once she admitted to herself that humble Omar with the Corolla was not a suitor her parents would accept. Marriage was a family affair in Afghanistan; he was an adventure that had to be left behind. There were tearful goodbyes on the eves of weddings, notes delivered by complicit sisters, hurtful texts; and then Omar would mourn, listening to "My Heart Will Go On," insatiably da capo.

A couple of years after his family moved into the rental house, his cellphone rang in the evening.

"Hello?"

"Salaam." He recognized the low timbre in her voice; it was Laila. She'd somehow gotten his number and borrowed a friend's cellphone, since she wasn't allowed her own. She had called to confess that she'd been in love since the first day they met in the alley. She wanted to be with him. She'd never had a boyfriend before, but she had friends who did, in secret—there was an underground of savvy girls at her high school, who traded boys' numbers like black-market commodities.

But Omar thought she was still too young for him—he was almost thirty then, twice her age. He didn't want to break her heart. So he told her the truth: He wasn't a good boy. He'd already had many girlfriends.

She said she didn't care, just like she didn't mind that his family was Sunni and hers was Shia, or that he didn't have money. She had once lived in a rental house, too, before the Americans came and her father got rich.

Thus began their secret phone conversations. Laila got her own cellphone, which she kept hidden in her room, and she'd call once her parents were asleep. Her father was a severely religious man who forbade his daughter to watch the television in his absence, locking it in its cupboard when he went out. Laila had never been allowed to watch Bollywood films, which meant she didn't know that she looked like Karisma Kapoor, but she did, Omar told her—the same glossy hair, petite stature, and teasing glances. He sang to her as Akshay Kumar sang to Karisma:

The world changes, the season changes, but my heart never will.

Laila hardly got to go anywhere besides her school, the mosque, and visits to relatives' houses. Her late-night calls with Omar were a window to the outside world. But though she loved to hear about his adventures around the country, and his dreams of a better life abroad, it wasn't as if she wanted him to take her away. She knew little of life, but already hers seemed filled by the prospect of a home and children. She found it hard to imagine leaving her family and country behind, and wasn't sure that things would be so much better elsewhere. Sometimes they stayed up past midnight debating these matters of the heart, but when Omar asked her to go to bed so that she wouldn't be tired for school, she'd plead for just five more minutes.

AFTER I MOVED INTO MY house in 2012, Omar invited me over to meet his mother, Maryam, who had offered to cook the national dish, *qabuli palao.* In Afghanistan it was an unusual gesture of intimacy to invite an unrelated male into the domestic sphere, but then again his mother was the head of the household. Omar's father lived in another part of town, and I assumed he and Maryam were divorced, although I didn't pry because I could tell the subject pained Omar.

He picked me up and as we drove the neat, paved grid of Qala-e Fatullah gave way to winding, muddy roads, until we arrived at his bazaar, where butchers hung their flyspecked wares outside, and teens tended cellphone shops. Beside the open gutters, hawkers called from carts laden with watermelons and Chinese brassieres; the smell of sewage and fried dough hung in the air.

I was sitting cross-legged on the carpet in the living room when his mother, Maryam, entered, a compact woman with a floral scarf thrown loosely over her dark hair. After we exchanged greetings, she turned to Omar. "So this is the foreigner I've been hearing about," she said, and laughed. "He does look Afghan."

Omar carried in the dishes: sliced cucumbers and onions with chilies, okra stewed with tomatoes, a stack of fresh *naan,* and the *qabuli*: a mound of rice laced with softened carrots and raisins. Beneath the long, shiny grains you could see dark hunks of fatty mutton. Plumes of rich vapor rose as I heaped my plate.

"*Nushi jan,*" commanded Maryam. "Eat, eat lots." A teacher in a government high school, she had the hale manner common among Afghan women who lead public lives. Maryam had been among the first generation of middle-class girls in her country who were educated for work outside the home. She'd been fortunate, she told me, that her father valued learning for his daughters as much as for his sons. He'd often cited a proverb to her and her sister: *Seek knowledge from cradle to grave.*

Maryam was fifty-four when we met, only a year older than my own mother, but within her lifetime she'd seen changes that had taken centuries in Europe, a leap from subsistence farming to the Internet. The students she was teaching faced troubles unknown in her childhood, such as broken families or heroin addiction. The children of returning refugees and rural migrants who'd flocked to the capital's relative wealth and security, they shared a discordant mix of manners and dialects. Afghanistan's urbanization, delayed by previous wars, was accelerated by this one, and since 2001, Kabul's population had doubled to more than four million. And in the city, you were connected to the world. Many of Maryam's students were as eager to learn as she had once been, but for this new generation, the desire for education was rivaled by that for emigration.

Nothing is intolerable until an alternative exists, if only as a dream. When Maryam, pregnant with Omar, had fled the Soviets, the limits of her horizon were Pakistan and Iran. Now her children were linked to a diaspora that stretched from Long Island to Melbourne, and the screens in their pockets brought them images of how life could be elsewhere.

People migrate because of the difference between here and there. Our world is divided between plenty and poverty, and just as a minority in each nation owns the majority of wealth, so too have rich countries consumed most of the planet's resources. More than half of global wealth is concentrated in North America and Europe, home to 15 percent of the population. Even after adjusting for the cost of living, per capita American income is thirty times that of Afghans. Economists refer to a *citizenship premium* which measures how much—all else, such as education, being equal—someone earns simply as a result of living in a particular country. It is as much as ten times more valuable to be the same individual in America or Europe than in a poor country; that is how much he or she might gain by crossing a border. Inequality is the slope of the frontier. It is the height of the wall that a person will scale.

Maryam had six children, but her eldest, Khalid, was the only one she gave birth to in her own country, in 1980. Omar was born two years later in Pakistan; then, after the family moved to Iran, came the third son, Mansoor, followed by her two girls, Haniya and Farah, and finally Zia, the baby boy. Perhaps their childhoods in exile had left them rootless but they all, like Omar, had dreamed of emigrating.

They knew that Afghanistan, one of the poorest countries in the world, was going to stay that way because of the war. But for her children to make it to better lives in the West, they would need six lucky breaks.

Khalid had been the first to go. Maryam's sister, a war widow, had been resettled as a refugee in Sweden in the 1990s. She came back to find a husband for her grown daughter, who was now a Swedish citizen. Marriages between first cousins were still common in Afghanistan— indeed, prized for the way they stitched families back together— and she asked Maryam for Khalid's hand. Khalid got married at the Swedish consulate in Pakistan, and was then issued a visa to fly to Stockholm. As she said goodbye to her eldest son, Maryam knew she wouldn't be permitted to visit him, but she was happy, nonetheless, for sometimes love meant wishing for separation.

Her youngest son, Zia, was the second child to emigrate. After dinner, Maryam asked Omar to bring their laptop, and he helped her find a video clip she wanted to play for me. Zia had filmed himself on a grainy webcam, lip-synching to a German pop tune. "He's become a European," Maryam said, chuckling wistfully.

Maryam was proud that Zia, unlike his older brothers, had never had to work as a child. He could focus on his studies and did well enough on his standardized English test to get into a university in the UK. The cost for his first year, including tuition, board, and flight, came to nearly ten thousand dollars. The whole family pitched in, including Omar with the money he was making on the front lines in Kandahar. The school had told Zia he'd probably get a scholarship the next year, but he didn't. He didn't have another ten thousand dollars and so, about to lose his visa and unwilling to return to Afghanistan, he went to Germany, ditched his passport, and applied for asylum.

When Maryam and I first met in 2012, Khaled and Zia were the only two of her children who had made it abroad. The others still lived with her in Kabul, where they had jobs working for the foreigners; by insisting her children study English, Maryam had prepared them well for Afghanistan's aid-driven economy. But by then it was clear that the money, like the troops, was on the way out—as were many of the elites who had most benefited from the American occupation.

Today, developing countries face not so much brain drain but wealth drain from outflows to real estate and bank accounts offshore. Before the US invasion connected Afghanistan to the global financial system, there wasn't that much a corrupt communist or Taliban official could do with their ill-gotten gains—invest in opium, maybe, or buy another car. Now billions flowed out legally as wealthy Afghans bought up property in places like Dubai and Malaysia and feathered their nests abroad.

For the rich, emigration is easy since citizenship, in the twenty-first century, is for sale. America's *immigrant investor* program requires nearly $2 million, while a *golden visa* to Greece needs only a quarter million euros. Afghanistan's speaker of parliament, who made his fortune supplying fuel to the foreigners, was reported to have secured EU passports for himself and his family in Cyprus for two million euros. With enough capital, you became a citizen of the world, what Frantz Fanon called *people without anchorage or horizon, colorless, stateless, rootless—a race of angels.*

The family's third chance came in 2014, the year that President Obama announced the end of America's combat mission in Afghanistan. Crime, unemployment, and suicide bombings were all getting worse in Kabul. One of Maryam's former colleagues who'd emigrated to Germany came back seeking a groom for her daughter. In Afghan culture, siblings are expected to marry in sequence, and to be skipped over is a source of shame. Many is the younger brother who pines for years, waiting for his elders. But Omar, who was next in line, wasn't interested. Nor did he care about tradition, he told me: Let Mansoor, his junior by two years, marry her. Mansoor was the better match anyway, dutiful and hardworking just like the eldest, Khalid, was. This time, the girl and her family flew to Kabul for the wedding. Omar congratulated his Europe-bound sibling without envy; he was still confident back then that he'd get the visa that America had promised for former interpreters. And when it came to marriage, he would follow his heart.

Several years after they began their friendship by phone, Laila had insisted on meeting, and he had finally agreed. She was eighteen now, a young woman getting ready for her first year in a religious college.

They started seeing each other for brief moments, at his house when no one was home, or in the attached garage at her place. Alone with her, he felt his blood kindle.

That was when I first learned about Laila. I was used to hearing about Omar's romantic escapades, and I didn't realize her importance at the time, but I remember clearly the harrowing story he told me one day, when I asked him why he looked so exhausted. The night before, he said, the neighbor's daughter had called him and asked him to come over. Laila's family was glued to their favorite TV show, and she was sure they wouldn't budge for the whole hour. Excusing herself to go do homework, she'd crept into the garage and let him in. They got into the back seat of the family car. In the gloom, with her breath in his ear, he felt the last restraints slipping—

Bang, bang, bang.

They frantically pulled on their clothes. Someone was knocking on the exterior door of the garage. In a moment, her father would come from the house to answer. Omar was trapped.

"Get under the car," Laila hissed. He threw himself down and slid underneath, as she climbed into the front seat and clicked on the stereo just as an arc of light stretched across the floor.

"What are you doing?" Her father stared at her from the doorway.

"I wanted to listen to the radio."

Omar watched the slippers pad around to the exterior door where, happy surprise, some relatives had dropped by. If the old man had any suspicions, he must have dismissed them, for they all went back into the house, leaving Omar underneath the car, his heart marking time against the concrete.

4

Worried that our plans to leave for Europe had been put on hold, I went over to Omar's house and asked Maryam what was going on with Laila's family. "Her age was just an excuse," Maryam told me, recounting how their landlord had rejected her suit. "The real reason is because we're Sunni, and they're Shia."

Omar wanted her to try again—in Afghanistan, a suitor's relatives are supposed to ask on his behalf—but Maryam was sure she would be refused. A proposal by his tenant's son must have alarmed the patriarch, who would now suspect his daughter of a secret liaison. For her part, Maryam didn't see what was so special about the girl or her family, who she thought were uneducated nouveau riche. She was worried that her son wouldn't leave Afghanistan. Along with Farah, her youngest daughter, and Suleyman, a teenage nephew that she'd fostered, she was getting ready to flee the country. Now that the borders were open in Europe, all she had to do was get to the Turkish coast and board one of the boats for the Greek islands. The sea crossing was dangerous but not long; from Athens, they'd be able to travel on buses and trains through the humanitarian corridor to Germany.

Maryam wanted to fly to Istanbul directly, but for that, they needed Turkish visas, which were expensive. As a civil servant, Maryam was able to apply directly, and paid only the official cost of sixty dollars. But most people had to go through a travel agency appointed by the

Turkish embassy, and, if they wanted their application to succeed, pay hefty, unofficial fees to middlemen—bribes, in essence. That fall, with so many Afghans desperate to reach Europe, the going rate was five thousand dollars for a simple tourist visa. It would exhaust her savings, but it was worth it to her to avoid the desert and mountains through Iran. She would even find money for Omar, but he refused to leave without getting engaged to Laila first. But couldn't he see how bad things were getting in Afghanistan? She told him that if the Taliban ever came to Kabul, they'd kill people like him who'd worked for the foreign military.

"He's gone crazy," Maryam told me. "Convince him this is his chance to go. You're his best friend."

"If he won't listen to his mother, what can I say?" I answered. "But I'll do what I can." As we parted, she gave me a complicit smile, and I wondered what she knew about my own plans with Omar, which were supposed to be secret. I had already told Maryam and the rest of the family that I wanted to write a book about Omar's journey to Europe, and I'd gotten their consent to tell their stories as well, but I hadn't said anything about going undercover as a refugee myself.

Like it or not, I was getting drawn into the whole family's westward migration, which had reached a critical juncture. Increasingly desperate in the face of their country's downward spiral, Maryam and her children had turned to what Afghans called the *rah-e qachaq*, or the smuggler's road. Haniya, the eldest daughter, had fled Afghanistan that way earlier that summer, just a few months before the border miracle.

Maryam had always felt guilty about bringing her girls back from Iran, where they'd grown up with far more freedom than they had here. Especially Haniya, who'd never been able to reconcile herself to Afghanistan's stiflingly patriarchal culture. Even if she wore the baggy cloaks she detested, she was still constantly harassed by men on the street, sometimes with actual propositions but more often with insults or furtive gropes. Her younger sister Farah tried to ignore it, but Haniya had a temper that could boil like her mother's. An avid soccer player, she'd also taken karate lessons in Iran. After the shopkeepers in the bazaar by their house had witnessed her chase down and slug a few men, they'd started tutting to one another: Madame Teacher is a respectable woman, but her daughter is *badmash*—a thug.

This was a country where a married woman was so subsumed into her husband's existence that it was shameful for strangers to use her given name; where people still said, approvingly, that so-and-so's daughter *hasn't yet seen the sun or moon*. Maryam had tried in vain to find her daughters a suitable match with an educated man who'd treat them with respect. So when Haniya, now twenty-five, told her that she wanted to seek asylum in Europe, Maryam assented, despite her fears.

When does a migrant become a refugee? Like her brothers, Haniya wanted to escape to a better life in the West. She'd lost hope in her own country, where the war was getting worse as the Taliban went on the offensive against the government: more than three hundred thousand Afghans were displaced by fighting in 2015, twice the previous year's total. And Afghan women faced additional violence: that spring, a young woman falsely accused of insulting the Quran was beaten and burned to death by a mob of men in downtown Kabul. A lone Afghan woman like Haniya was very likely to win asylum if she applied in Europe or Canada. But she had no legal way of getting there. A cordon of visas and laws regulating air travel existed to keep her and other refugees out. *This catch-22 for refugees*, as David Scott FitzGerald called it, has its antecedents in efforts to keep out Jews fleeing the Nazis, and it ensures that the more likely a person is to seek asylum in the West, the less likely they are able to board a flight there. Germany introduced visa requirements for Afghans in 1980, a year after the Soviet invasion. Afghanistan had one of the worst passports in the world when it came to visa-free travel.

Haniya knew she couldn't fly to Europe. She wasn't rich enough to buy residency, and there was no one with a Western passport who wanted to marry her. So she was the first of Maryam's children to take the smuggler's road. Europe's borders were still closed then. To save money, instead of getting a Turkish visa, Haniya paid a few hundred dollars for an Iranian one and flew to Tehran, where she met up with a male cousin and his family. They traveled with smugglers on a grueling journey over the mountains into Turkey, and then crossed into Bulgaria. Eventually Haniya reached Germany, where her brother picked her up. Exhausted but triumphant, she warned her mother, who had bad knees and diabetes, not to attempt the same overland journey from Iran.

Before Haniya's flight to Tehran, when Maryam had accompanied her to the airport to say goodbye, the eldest daughter had stopped at the security checkpoint and pulled out a brand-new pair of shoes.

"Take these, Mother," she said, handing Maryam her old ones. "I don't want to bring any of this country's dirt with me."

THERE WAS SOMEONE ELSE BESIDES Laila that Omar worried about leaving behind in Kabul. His father, Jamal, long absent from their lives, had returned to live with them a couple of years earlier. He and Maryam were not in fact divorced, as I had assumed, though they were certainly estranged. She told me she didn't care what became of her husband, who was refusing to go into exile again, even as the rest of the family fled abroad. But Omar didn't want to abandon him, as his father had once abandoned them.

I had first met Jamal over dinner a couple years earlier, soon after he came to live at the rental house. He was even taller than Omar, with the same thick eyebrows and square chin, but badly hobbled by the herniated disks in his back. Jamal was courteous to me, but we'd sat in awkward silence through dinner, Maryam visibly tense as she came in and out of the kitchen. As soon as we finished eating, Jamal had retired to his room.

Ever since Omar and his siblings were old enough to remember, their parents were always fighting; and even their moments of joy, like a birthday party or a visit to the park, were overshadowed by the fear they felt living as refugees on the margins of Iranian society. How much of his parents' strife was due to forced migration and poverty, and how much was the result of basic incompatibility, no one could say. Pain was pain. Yet they had stayed together for the children's sake through the bitter decades abroad, and it was only after they came back to Kabul in 2002 that they finally split. The family had gone to stay at a house that belonged to Jamal's brother, who had emigrated to America. Jamal's eldest sister had lived there throughout the war, and she'd recently married a much younger man. It was a big house, with six bedrooms, but it wasn't big enough. The new brother-in-law thought the house came with the spinster he'd wed. Maryam and the kids felt they had just as much of a right to it. Jamal had tried to mediate, but things soon got ugly.

The fight began over the pump at the well in their yard. The new husband yelled at Omar not to swing the handle so hard; when Omar talked back, the man came and knocked him flat; then his older brother Khalid broke a stick over the guy's back. When the police showed up, they arrested the two boys, along with Jamal for good measure, since he was their father. The next day, they were released, but the boys were ordered to leave the house. Maryam and her children packed their few possessions; they had little money and nowhere to go. Jamal had wrenched his bad back trying to break up the scuffle and was lying down in a bedroom; the littlest girl, Farah, went in to see him. She stood in the doorway, staring at her father's humped form, his face turned away from her.

"Father, are you coming?" she said at last.

Jamal didn't answer. He had enough of Maryam and the boys' disrespect; he was staying with his sister. And so he let his family go. Their separation would last more than a decade.

IF YOU ASKED THEM, BOTH Maryam and Jamal would insist that they never had a happy day together since they were married in the fall of 1979, in a modest ceremony in downtown Kabul. Yet Maryam's favorite teacher, an elderly woman now living in California, remembered her looking radiant. No photograph survives but it's easy to picture the couple on that day: Jamal broad-shouldered with a dashing mustache, Maryam petite beside him, adorned with jewelry. I like to imagine them happy in the beginning, or at least hopeful; it's saddest to have never lost anything at all. Perhaps their love would have grown if their homeland had remained at peace, as it had during their childhoods, when the nation was enchanted with *taraqqi*, progress, and when Kabul came alive with marvels like electric lights and cinemas. On Independence Day, the crowds had danced to Ahmad Zahir: *Life will sing in the end, slavery be gone!* The president, Mohammad Daud Khan, had embraced the third-way nationalism of Nasser and Nehru; he was a man happiest, it was said, *lighting an American cigarette with a Russian match*. After she graduated from college, Maryam was content to remain single, supporting her widowed mother with a job at the ministry of finance. On April 27, 1978, she'd been at work as usual when gunfire erupted and the big windows shattered. She and other girls escaped to their homes, where Radio Afghanistan gave them the news of a coup d'état: Nur Mohammad Taraki, leader of the communists, soon to be *True Son of the People and Chief Commander of the Great Saur Revolution*, was president. Two months later, Zahir, the voice of their generation, was killed in a car crash.

Maryam's father had passed away when she was a girl, and a household without a male protector was exposed to social pressure. For years, she'd fended off her maternal uncles, who wanted Maryam for their own sons, as if she would ever return to the village to live among livestock. But she was almost twenty at the time of the coup, well past the age when most girls were married off. People gossiped that she was *left at home*; each year would bring more pity and scorn. Jamal's brother was her coworker but she'd never met her future husband when his family proposed to her mother, after the communists took over. Jamal was raised by his two elder brothers, who wanted to marry him off so he'd stop coming home late. From Jamal's photo, Maryam could tell he was handsome. They were married that November. A month later, the Soviets invaded.

For the next two years, Maryam and Jamal clung to their life in Kabul as the war intensified. Like the Americans in Vietnam, the Soviets used bombing campaigns to displace people from the countryside and flush out the guerillas. The worst of the war was in the rural areas, but the mujahideen set off bombs in the capital, while the communists hunted for spies and sympathizers. Two of Maryam's cousins from her nearby village were arrested and were never seen again. When the regime began conscripting men like Jamal who'd already done their service, they decided to flee, crossing through the mountains with smugglers to Pakistan. Their firstborn, Khalid, was a year old and Maryam was six months pregnant with her next child. In the spring of 1982, in a hotel in muggy Peshawar, she gave birth to Omar.

When they ran out of money, they had to go into camps set up for Afghan refugees. They were assigned to one in a steep and beautiful valley in Mansehra. Maryam despaired when she saw the women lugging water they fetched from the river below. It was the village drudgery she'd always feared. With their beards and turbans, the men seemed frighteningly tribal to her; some had distant, hollow stares, having come back recently from fighting the Soviets. To live in the camps and receive UN aid, Afghans had to register with one of the seven mujahideen parties supported by Pakistan. *Inside these camps was a huge reservoir of potential recruits for the Jehad*, wrote a Pakistani brigadier. *Thousands of young boys came to the camps as refugees, grew up, and then followed their fathers and brothers to the war.*

During that decade, more than six million people would flee across the border to Iran and Pakistan, forming the largest group of refugees in the world, a distinction Afghans would hold for the next thirty years.

FROM THE BEGINNING, THE MODERN idea of a refugee has been tangled up with power and politics. In 1951, when a group of delegates gathered in Geneva to draft a treaty on refugees for the UN, none of the communist countries participated. The debate on the shores of Lake Leman turned around a central question: Who was a refugee? In contrast to other migrants, this foreigner would be granted many of the same rights as a citizen; most fundamentally, the right to remain. A refugee could not be deported back to danger. Signatory states would be asked to accept an unprecedented limit to their own sovereignty. With the cataclysmic displacements of the Second World War fresh on everyone's mind, some delegates argued for a broad definition that would include those escaping violence in general, but the United States was keen to limit it to anti-communist uses. Thus according to the 1951 Geneva Convention, the foundation of international refugee law, a refugee is someone with *a well-founded fear of being persecuted for reasons of race, religion, nationality, membership of a particular social group, or political opinion*, and not someone simply fleeing war or disaster—criteria tailored to the Cold War dissident.

When it came to those escaping communism, the West could be generous. In 1956, as Hungarians fled the Soviet occupation, the American public clamored for them to be given asylum. When Saigon fell to the People's Army, 140,000 Vietnamese allies and their families were evacuated and brought to the United States. After the war, people continued to escape in craft ranging from small wooden fishing boats to steel freighters chartered by smuggling groups. By June 1979, nearly 54,000 *boat people*, as they were dubbed by the Western media, were landing each month in Southeast Asia, most Vietnamese but some fleeing communist regimes in Laos and Cambodia. Local authorities began forcing refugee boats back at gunpoint; pirates raped and murdered thousands. At a conference in 1979, the West struck a deal with Vietnam's neighbors: *an open shore for an open door*. In exchange for regional countries granting temporary asylum to the boat people, the West committed to resettling them. In the next three years, more than six hundred thousand refugees emigrated, the vast majority to the United States, Canada, and Australia.

Broadcast on TV screens in the West, the plight of the boat people touched consciences already guilty over the war in Vietnam. A new humanitarian politics was emerging that would scramble the old ideological spectrum and outlast the Cold War. *It's very simple*, said the writer Heinrich Böll, about his privately funded ship that rescued boat people and resettled them in West Germany. *I am of the opinion that people should save lives where they can be saved.*

As the world got smaller, people were visible to one another in new ways, and the sight of destitute strangers did not always inspire pity. As the developing world's population boomed, migration was increasing from the former colonies to Europe and North America. The backlash to this in the West was, in a sense, the inverse of human-itarianism; both were emotional reactions to the figure of the Other, whose face is, according to Emmanuel Levinas, *at once the temptation to kill and the call to peace.*

In his notorious novel *The Camp of the Saints*, published in France in 1973, Jean Raspail imagined a vast fleet of migrants, *chased by fam-ine, misery, and misfortune*, setting sail for Europe. *The Third World overflowed*, Raspail wrote, *and the West was its sewer.* Often dismissed for his crass racism, Raspail's insight was that politics was becoming a struggle over the sentiments of the rich for the poor. Enfeebled by hu-manitarian pieties, and by the *little masterpieces of indignation* about global poverty that appeared alongside *luxury advertisements* in the press, the West, Raspail believed, had *lost the force and will to say, No!*

In the book, the French president, foreseeing the destruction of his nation, sends a warship to intercept the migrant flotilla—but the cap-tain reports back that his crew, overcome by pity upon seeing women and children, refuses to open fire. France faces a stark choice, the offi-cer tells the president: *We either take these people in, or torpedo every one of their boats at night, so we can't see their faces as we murder them.*

WHEN I WAS A BOY, my father told me a story about the little boats. In the summer of 1990, he was an officer aboard the *Huron*, a Canadian navy destroyer sailing the South China Sea, when they saw a vessel in the distance. It was a wooden junk with eyes painted on the prow, about sixty feet long. The open deck was crowded with people in soiled clothing. They'd escaped Vietnam several weeks earlier, and were headed for Malaysia, when a storm knocked out their engine. Adrift outside the shipping lanes under the blazing sun, they were dying of dehydration and dysentery, fifteen already gone. They might have be-come another of the sea's mysteries, but late in the day a Canadian helicopter had spotted them, and the squadron came to their rescue.

There's a video that someone uploaded to YouTube filmed from the *Huron's* bridge, where my father was standing. The camera zooms in on the boat's aft deck, where a boy lies facedown and motionless.

The refugees crane their faces upward, smiling and pleading with their hands joined in prayer. Between bursts of radio traffic, you can hear their cries for help carry across the water. Ninety people were taken aboard the squadron's supply ship. Two died and were buried at sea; the others were offered asylum in Canada.

My father kept photos of his four black-haired kids tucked inside his cap while he was away for months at sea. I remember how the diesel smelled in his uniform when he came back, and the feel of brass buttons and wool against my cheek. He was sailing the Sea of Japan the day my mother gave birth to me in British Columbia and he only met his firstborn three months later. I was waiting dockside in my mother's arms, and a newspaper photographer took a picture of the three of us. Later, someone clipped it out and mailed it to him, scrawling across it: *Another Metis bastard is born.*

My ancestors were migrants, like those of most North Americans, but they crossed different oceans. Aikins is a Scottish surname by way of Ireland, and there's *pure laine* Quebecois on my father's side as well. His folks were among the fifty-five million Europeans who emigrated from the nineteenth century onward, when the first industrial revolution mixed the peoples of the earth by ship and train. My middle name is Yutaka, 豊 in kanji, after my grandfather, who was born in California in 1922, one of the nisei, the second generation of Japanese in America. His father was a merchant seaman who jumped ship in San Francisco; both my great-grandfathers labored on farms in the Sacramento Valley and married *picture brides* from Wakayama and Hiroshima, bringing them over before Asians were banned under the Immigration Act of 1924, the same year the Border Patrol was created. The act was passed after decades of lobbying by groups like the Native Sons of the Golden West and the American Legion. *Anyone who has traveled in the Far East knows that the mingling of Asiatic blood with European or American blood produces, in nine cases out of ten, the most unfortunate results,* wrote Franklin Delano Roosevelt in the *Macon Telegraph* the following year. When the Japanese attacked Pearl Harbor, Roosevelt was president, and in response he authorized the internment of more than a hundred thousand people of Japanese descent living on the West Coast, two-thirds of them American citizens. *They are always going to be brown men,* said Senator Edwin Johnson. *Do you think they will finally merge and just be accepted in every way like a white man?*

My grandmother Sei, later known as Mary Ann, grew up outside Los Angeles; she was thirteen when war broke out. On the school bus, some white kids spit on her because she had the face of the enemy. When the government ordered Japanese Americans to report to camps, they were told to bring what they could carry, and lost almost everything else. Sei and her family went to the *relocation center* at the Santa Anita racetrack, where they were put in whitewashed horse stalls. When her father balked at answering the *loyalty questionnaire*, they were taken to the *segregation camp* at Tule Lake until he changed his mind. Then they were transferred to swampland in Arkansas, near the Mississippi River. It was 1943 and Japanese Americans were being shipped to an inland archipelago of tar-paper shacks and barbed wire, places like Topaz, Gila River, and Heart Mountain. My grandfather Yutaka's family was sent to Camp Amache, out in Colorado, which took its name from a local Cheyenne woman who married a white rancher and whose father, Lone Bear, was among those massacred by the army at Sand Creek.

What has happened lately to the American Japanese and what has happened all along to us, wrote Langston Hughes in 1944, *puts American Negroes and American Japanese in the same boat.* Even after they got their freedom back, many nisei stayed away from the West Coast, fanning out on trains and Greyhound buses to factories, irrigation projects, and new suburbs. My grandparents met in Denver on the assembly line of the Acme Table Pad Company: he was seven years older with one failed marriage behind him, a charmer with a weakness for Chinese gambling, someone she fell for in spite of her parents' disapproval. They both wanted something better than factory labor or housework, so my grandparents kept moving onward, landing in Chicago, where they got their break: chick-sexing. At the time, Japanese Americans had a near monopoly on a remarkably efficient method of determining the sex of baby chickens, invented in Japan in the 1920s. Chick sexers, through constant repetition, would develop an intuitive ability to recognize the chicks' genitals, which were indistinguishable to novices. A skilled chick sexer, who could *throw* more than a thousand chicks an hour with 98 percent accuracy, made good money. When a sexing franchise became available in Mississippi, my grandparents moved to the state capital, Jackson, where their fifth daughter, my mother, was born in 1959. Those were the dying years of Jim Crow, and whites in Mississippi were fighting to keep their schools segregated, but my grandparents were sold a tract house in a white neighborhood and their daughters were allowed into school there without a fuss. By then, white veterans were coming back from overseas with Japanese and Korean brides.

Yutaka had loved cars since he was a boy; to him they meant liberty. One day he parked a big-rig truck on the lawn, which he'd bought with the family savings. He was done with chickens; the future was refrigerated shipping. After that he was seldom home, driving and smoking Winstons to stay awake, trying to make the bet pay off. The summer of 1964—*Freedom Summer*, the volunteers coming to Mississippi from across America called it—Yutaka was running the freeways to California and Illinois, Highway 51 from New Orleans to Memphis, dirt roads in the Delta.

On September 22, Yutaka filed for bankruptcy at the courthouse in Jackson. That same afternoon, he went out to Indianola to unload the truck and dropped dead of a heart attack. He was forty-two; the eldest of his seven children was twelve, the youngest two. My grandmother had to take care of bankruptcy and funeral arrangements all at once. She moved to Seattle, where she had family; ran a flower shop and never remarried. Survival for her, like many nisei, meant becoming fully American, as if the government's questionnaire was in the back of her mind: *Will you avoid the use of the Japanese language except when necessary?* She'd given her children Christian names and made sure they went to college. All but one married white spouses.

My mother met a young naval officer, my father, after his ship made a port call in Seattle, and went away with him to Canada, to suburbs where my siblings and I played ice hockey. I remember isolated incidents, a woman using a slur against my mother over a parking spot; a bully at school; and I was aware that I looked different, less desirable, but inside I felt the same as my white friends. My grandmother died when I was in college, and I never asked her about the camps. As a boy, the history that interested me was on my father's shelves: Winston Churchill's memoirs and *The White and the Gold*, a saga of French colonial heroes. As I came of age, I worked my way up through progressively cosmopolitan spheres, reaching the apex of New York City, the beneficiary of what the historian Nell Irvin Painter called *the fourth enlargement of American whiteness*. Today a multicultural class has, following propertyless Englishmen, Irish and Germans, then Italians and Jews, inherited an America where Thien Thanh Thi Nguyen, the daughter of Vietnamese refugees, can grow up to be Tila Tequila, famous as the first bisexual reality TV star and, more recently, a neo-Nazi.

WHEN OMAR WAS LITTLE, HIS family's favorite show was *Road to Avonlea*, a dubbed Canadian period drama that was popular in Iran, where the family had moved when he was still a toddler. The main character, Sara Stanley, had hair as golden as the hay fields of Prince Edward Island; she was a city girl exiled and sent to live with her country relatives. Sara, Sara, Sara, with her gray cat's eyes; he swore to his amused family that, when he got older, he would go to Canada and marry her.

They lived in Shiraz, where the poets Hafez and Saadi are buried. They were eight in a little house in the old part of the city. Unlike in Pakistan, where Afghans were mostly kept in camps, in Iran, they were allowed into the cities where their labor was needed, as local men left for the front lines of the war raging with Iraq. Times were nonetheless lean for the family, especially after Jamal threw out his back working on a masonry crew. But in Iran it was easier for women to work outside the home. Maryam cleaned houses and ironed the secondhand clothes that her neighbor sold in the bazaar. She sometimes worked the *suitcase trade* from the port of Bandar Abbas, bringing in black-market goods in her luggage, which little Omar and his brothers hawked on the side of the road, along with Minoo Biscuits and Chic Gum.

Even with marital bliss, it would have been tough going, but Jamal and Maryam had never found harmony. According to tradition, the woman was supposed to yield, and Maryam would not. Jamal, laid up at home and jobless, grew angry that his wife would not heed his authority, even when he struck her and their fights grew wild. Many years later, he would wonder to me whether he should have used his full force, until something broke. Before they fled Kabul, his older sister had taunted him sometimes over Maryam's willfulness, saying, *Your father had many wives and you can't handle one*, but here in Shiraz he was alone.

As he grew up, Omar's looks came to resemble Jamal's more than any of the boys. He shared his father's musical tastes, too, and loved to listen to his cassettes: oldies like Ahmad Zahir but also a new generation of diaspora singers, voices from the West that longed for home, such as Farhad Darya, whose song "Kabul Jan" was a hit in the nineties:

> *Let me sing from Iran to Pakistan.*
> *Do you have news from dear Kabul, beloved?*

Afghans were increasingly unwelcome in Iran, and the police became more predatory. But Omar, with his light eyes, had it easier than the Hazaras who were marked by their more Asian faces, who got followed home by bullies calling: "*Kesofat-e afghani!* Afghan garbage!"

No, his hair was brown and his accent sounded like any Shirazi's. But his swagger was modeled on the angry young men in the Bollywood films he adored. He was big for his age, and hot-blooded. He learned how to handle a knife from the local toughs, the *badmash*, but stayed clear of the heroin that was raging through the city.

He kept working right through school, enough to buy himself a bicycle and a radio. He shined shoes and harvested pistachios. At the age of ten, he could lift fifty-pound tubs of ice cream at the shop. At thirteen, he started working in construction. When he grew up, he wanted to become a doctor, and help people. Or a pilot, and see the whole world. He knew there was a better life somewhere, in a place where he would belong. A friend of his lived near the airport, and they'd go out and spend hours by the fence, watching for planes.

THE END OF THE COLD War changed how the West saw refugees. Conflicts in poor countries were no longer deemed proxy battles between superpowers and their ideologies. *Terrible wars, without faith or law, no less foreign to the logic of Clausewitz than to that of Hegel,* wrote the war-correspondent-cum-philosopher Bernard -Henri Lévy.

After the Soviets left Afghanistan in 1989, millions of Afghans went home. Jamal wanted to return, but Maryam refused. She didn't trust the fundamentalist rebels to let her girls continue their education. When the communist government collapsed three years later, the mujahideen and militias turned on each other, and the West turned away. *The world has no business in that country's tribal disputes and blood feuds,* proclaimed the *Times* of London.

Now came the time of the warlords. *Love died, devotion died, affection perished,* the poet Khalilullah Khalili had written. *We are drowning in a sea of blood.* The devastation of the nineties would stand as a warning to Afghans two decades later, when the Americans started to go and the government teetered: Flee while you can. Kabul became a maze of ruins carved up by armed checkpoints. Crossing them meant risking death; it might be the only way to avoid starvation. From a letter, Maryam learned that some militiamen had robbed her brother and beat him until he went blind in one eye. Then the Taliban, with harsh justice as their almost singular attraction, swept across the country. In 1996, they captured Kabul.

Five years later, Omar's family received long-awaited good news: Jamal's brother, who'd emigrated to the United States a decade earlier, was going to sponsor their refugee application there. But they had to leave Iran and go to Pakistan, where there was an American consulate. They sold what they couldn't carry and went to stay with one of Jamal's sisters in Peshawar. When Maryam and Jamal went to apply at the US embassy, they were scheduled for a follow-up interview a few months later. It was late summer, 2001.

On September 11, they watched on TV as the Twin Towers burned. The US Special Forces and their warlord allies were in Kabul by the time they returned to the embassy. The official there told them that refugee applications from Afghans were being suspended. Maryam recalled his words: "Your country is free now. Go home."

THEY RETURNED IN BATTERED TOYOTA pickups and Mercedes buses long retired from Germany. They came on foot and on donkey, family by family over mountain passes, and in throngs of thousands at the great border crossings of Wesh and Torkham. They landed in Boeing 747s, wearing suits and freshly shined shoes. They came bearing master's degrees, and prosthetic limbs, and red-cheeked infants.

Five million returned. It was the largest repatriation program in UN history. *Our commitment to a stable and free and peaceful Afghanistan is a long-term commitment,* said President George Bush, standing beside the new Afghan leader, Hamid Karzai. The Taliban had prohibited music; the first song played by Radio Afghanistan was Darya's "Kabul Jan":

> *Do you have news from dear Kabul, beloved?*
> *And have you heard from me?*

In the spring of 2002, Omar and his family took a bus up the Khyber Pass and arrived at the border of their own country. At the crossing, the stark white banner of the Taliban had been replaced by the black, red, and green of the king's time. The sight of the flag filled Omar with hope.

5

Miracles never last.

The winter came and went while Maryam and the others waited on their Turkish visas. I finished the Kunduz story and wrapped up some loose ends: we found Turabaz, our old *chowkidar*, a job at an NGO, and Baad the dog went to a farm north of the city. I put the rest of my life on hold and immersed myself in preparations for the trip.

Then came bad news from Europe. First, the borders were closed in the Balkans. Next, on March 18, the EU announced that any refugees who landed on the Greek islands would be forced to stay there in camps. Our journey had gotten a lot harder—though I was beginning to wonder if Omar would ever leave.

"Spring is here," I said, as we sat in the Corolla outside the place I was staying. "What are you going to do?"

He wasn't listening. He was fixated on the thought of Laila's father marrying her off once he left. As long as he was still in Kabul, he'd be able to prevent it—by force, if necessary.

"What do you mean, by force?" I asked him.

He laughed bitterly. "I can prove that we were together. No one else would marry her then. Or—I'll kidnap her."

I couldn't tell if he was serious. I'd never seen him like this.

"What does she say?"

Laila's father had searched her room and found her cellphone;

now Omar had to wait days at a time for her to reach him through a classmate.

"She says that if I leave, and her father forces her to marry someone else, then I can't complain."

Not for the first time, I wondered what Laila was really thinking. How would they marry if she was unwilling to defy her father? Omar had told me that she'd finally accepted that it was best to leave, but perhaps she was really trying to keep him here. I'd given up trying to argue with him. Maybe I was just wasting my time.

Omar told me he'd find a way to win her. He'd get rich somehow. He was willing to do anything apart from kill or steal. He would buy hashish in Ghazni or Kandahar, and sell it in Kabul for twice the price.

I told him he was crazy. He agreed.

"I'm crazy with love. I've been with her for four years. I was the first person she was with. I can't stand the idea of anyone else having her."

"You know, there's an expression in English: 'If you love someone, set them free.'"

"Give them up?" he asked.

"And if they come back, then they love you."

"I can't give her up, brother, I can't."

"But that's not love."

"It's not?"

"It's pride."

"I know I love her."

SHE TOLD ME THAT SINCE *death is in the air here*, wrote the poet Elyas Alavi, *you could shut every little door and in the end it would still come into your room.*

Shortly before nine o'clock on the morning of April 19, I was sitting at my desk in my friend's house, practicing Dari, when a clap of wind swung the window inward. Car alarms and barking dogs resounded in the street. I checked Twitter; soon people were posting pictures of a dark bolus of smoke rising from the center of town. There'd been an explosion at Mahmood Khan Bridge; a big one, guessing by its deep boom. I called Omar to come over, hopped in the Corolla, and we made our way downtown through the traffic, laden ambulances rushing in the opposite direction, passing the billboard by the German embassy that read: LEAVING AFGHANISTAN? ARE YOU SURE?

"Look at them. They couldn't care less," Omar remarked as he threaded the car past some laborers shoveling gravel. People got on with their lives, even as the bombers walked among them. But the blasts were getting bigger, as the foreigners and elites swaddled themselves in concrete walls and checkpoints. And now that ISIS had set up a franchise in Afghanistan, Shia schools and mosques were being targeted. That year, the UN would record the highest toll yet of civilian casualties in Afghanistan, with more than eleven thousand killed or injured, a third of them children.

The police had set up an outer cordon around the blast site, so Omar dropped me off. I flashed my press badge and walked on through the streets, deserted except for a few dazed-looking shopkeepers sweeping broken glass. It started to rain, soaking the loafers I'd forgotten to change. I could hear heavy gunfire up ahead. Passing through a warren of auto-parts shops, I arrived at a vast yard strewn with corrugated-metal shacks. The fighting was still going on across the road, where an intelligence compound had been hit. Inside one of the mechanics' shacks, I found a group of Afghan reporters huddled and waiting. The rain drummed on the roof and it smelled like wet polyester suits. I spotted Massoud Hossaini, who'd won the Pulitzer Prize four years earlier for his photo of a little girl, her green Ashura dress spattered with blood, screaming as she looked down at the bodies of her family.

Today's terrorist attack near Pul-e Mahmud Khan, Kabul clearly shows the enemy's defeat in face-to-face battle, the office of the president, Ashraf Ghani, tweeted. The Taliban had detonated a van full of explosives against the compound wall; afterward, three armed men in stolen uniforms charged in, shooting the survivors. The shock wave swept through a nearby roundabout and marketplace. Almost seventy people were killed in that day's bombing, more than three hundred fifty wounded, a mix of security forces and civilians. Some people survived miraculously, like the bicycle repairer who was thrown clear into the river, swollen with spring rain. Others simply vanished. When I returned to the ruined market a week later, a woman was wailing for her son: "My child, my child, I've gone blind. What's become of your flesh?"

"I HEARD SOMETHING STRANGE LAST night," Omar told me. He'd been alone in the house with his father, whom Omar and his brothers in Europe had been debating what to do with. None of the siblings wanted to leave the old man behind but since Omar was still in Kabul the burden had fallen on his shoulders. He'd always been the son who was closest to the father, anyhow. After Jamal abandoned them in 2002, Omar, along with Farah, the youngest girl, would visit him sometimes at the aunt's house. Despite his hurt and anger, Omar still loved his father. The big man who'd once ruled them with his belt and fists was growing frail. When his father underwent surgery for his bad back, Omar drove him to the hospital. When the aunt died, and the uncle in America decided to sell the house, it had been Omar and Farah who went over and asked Jamal to come live with them. It was shameful to have their father staying like a beggar in other people's homes. At first Jamal didn't believe them. Was it a trick? But he had nowhere else to go.

Neither did Maryam, who grudgingly accepted the presence of her estranged husband. With small humiliations, like serving him last, she did her best to make him feel unwanted. And the children didn't respect their father's authority anymore. Jamal took to fixing his own meals and using the outhouse in the yard, even in winter, and spent the rest of the time alone in his room, watching television, smoking cigarettes.

But now his whole family was escaping to Europe. He'd sworn never to leave his own country again. He threatened to go to Kandahar if they left him on his own, and that would be the last anyone would hear of him. Omar thought it was a bluff; his brothers weren't so sure. But even if their father was willing to fly to Turkey with the others, would they be able to scrape together five thousand dollars for his visa? The old man was penniless.

The previous night, Omar told me, his father must have thought the house was empty. Omar sat in his room, listening to a sound he'd never heard before. The old man was crying.

THE TURKISH VISAS HAD COME at last, and would soon expire. Maryam waited as long as she could for Omar. Then in May, when the summer blossoms arrived, she asked her son to drive her to her ancestral village outside the city. Suleyman, her fourteen-year-old nephew who was accompanying her to Europe, joined as well, so that he could say goodbye to his parents and siblings, who still lived in the village. Omar took the road along the far side of the airport, past the idled cement yards with their samples of blast walls and pillboxes set out like garden gnomes. At the intersections, children sold bitter melons.

When she was a child, Maryam lived in Nangahar, where her father worked as a civil servant, but in the summer her mother would take her and her siblings back to her home village outside Kabul. Maryam's maternal uncles tilled their wheat fields with wooden plows; within the mud walls of their compounds, the women tended the grape orchards and made bread from flour ground at the water mill. Maryam was young enough to run free with her male cousins through the fields, climbing the banks of canals full of clean water, warm in the summer and cool in the fall, when the vines grew heavy with dusky fruit. The climate was less arid back then.

Omar drove past the kinked strand of concertina that marked the checkpoint on the edge of the Shomali Plain, an agricultural valley that stretched forty miles north, past Bagram Air Base. During the surge years, there had been a grand vision for a new Kabul here, designed by French and Japanese architects, where three million people would live in modern housing with water and electricity. But nothing had come of it, and the plain beyond the highway was bare except for the funnels of brick factories worked by rural migrants. Paid for each thousandth brick, children, their parents, and grandparents toiled side by side from morning to night, molding the earth in wooden frames and firing kilns with shredded tires that gave off dioxins and heavy metals.

Omar followed the curving road northwest for a couple of miles until he reached the village's fields and orchards, bisected by earth walls. Up on the ridge, green flags fluttered above gravestones; Maryam's mother was buried there. Omar wheeled the Corolla onto a dirt road, past the elementary school and through a cluster of houses along the riverbed, where children stood in the shade cast by a mulberry copse. They passed by the sawtoothed walls of an archaic mud fort, dilapidated by Soviet shells and erosion. It was the abandoned *qala* of Maryam's grandfather, said to be haunted. Today, his descendants from three wives, numbering some three hundred, owned much of the land. Part of it, Omar once told me bitterly, should have been Maryam's, from her mother's half share mandated by Islamic law. But her kin had taken it, saying Maryam had forfeited it by marrying outside the village, and by going abroad during the war.

Omar dropped them off at the house of Suleyman's father, Ismail. A spry, bowlegged farmer in his late fifties, Ismail was Maryam's favorite cousin. Her mother had come to stay with him when the warlords ravaged the capital in the nineties, and she had passed away in this home, while Maryam was living in Iran. Her cousin Ismail had sent Suleyman, his eldest son, to live with Maryam in Kabul so that the boy could attend a proper school there, and when the farmer heard that she was leaving for Europe, he'd scraped together the money to buy his son a visa. "You can study there," he told him.

More than a hundred people came during the week Maryam spent in the village. Her relatives knew that this might be the last time they saw her. They respected her as an *adam-e mardana*, a manly person, who'd become a teacher and, despite a rotten husband and worse luck, had raised all six of her children to adulthood. Together they sang songs and gossiped over meals and recalled the many who were absent. Her generation's life expectancy had been about thirty years. To say you miss someone in Persian is to say *your place is empty*. She missed her cousins, Ismail's older brothers, the boys she played with in the time of grapes and wheat. The communists took them and they disappeared. But two of their children, cousins to each other, were now married, so the two brothers would have had the same grandchildren.

The community gave individual life meaning here, but even as a girl, Maryam had not wanted to marry a village cousin. She had wished to be modern. Her children had never even considered such a life; one after the other, they had gone into exile in Western cities.

Each day, Maryam went to pray by her mother's grave: *In the name of God, most merciful and compassionate*. Here the dead were still present for the living, and Maryam wondered if she would be separated from them, too, in Europe. *You alone we worship, and you alone we ask for help. Guide us on the straight path.* A solitary woman on a barren hillside. Was she weeping for what she'd already lost or what she would leave behind?

After seven days, Omar came to fetch her. Maryam embraced her kin, and they asked one another for forgiveness in this life and the next.

IN THE MORNINGS, HE DROVE past Laila's college, hoping to glimpse her face. He'd been expecting her to call for weeks. Finally, he waited outside her house and followed her to school. When she got out of the taxi and was walking toward the entrance with a friend, he pulled up.

"Hey!" he called out the window. "I love you."

She ignored him. At the entrance, the friend went inside, while she continued toward a shop.

He cried to her again: "I love you!"

This time she wheeled on him and cursed: "Stop following me, honorless dog!"

She went into the store. He parked the car. When she came out, he seized her by the shoulders: "Why do you speak to me like that? Why don't you ever call or talk to me? I love you, don't you know that? Didn't you say you loved me? What happened? What's going on? Please, I'm going crazy!"

She had turned pale, and only said, softly: "Please let me go. The people are watching."

And he did.

"MY POOR MOTHER," OMAR MURMURED. He told me that, as he was saying goodbye to Maryam at the airport, she'd handed him an envelope stuffed with US dollars, money from her savings, and that his brothers in Europe had sent.

"For your father," Maryam said. That, plus another thousand from Omar's own pocket, was enough to buy Jamal a Turkish visa. The old man's bluster about running away evaporated, and Omar took him for passport photos and a visa interview. His father would fly to Istanbul and join the family there, to try to find a way into Europe.

The night before his father's departure, Omar and I brought some takeout kebab to his house. As we were pulling in, Jamal showed up on his creaky old bicycle. We politely invited him to join us and, to my surprise, he accepted.

We sat cross-legged in the living room and scooped up chunks of *chapli* kebab with our bread. The old man peered at me from under his silvery brow. "You've traveled the whole world," he said, breaking the silence. "Tell me, which countries were the best, and which the worst?"

"Each has its good and bad points, I guess," I answered.

"What do you think about Afghanistan?"

"It's very poor and the people are suffering, but they take care of each other." I thought for a moment. "And it's one of the most beautiful countries I've ever seen."

He brought a fizzing glass of cola to his lips. "Have you seen the Golden Gate Bridge?"

"I have."

"Is it still the longest in the world?"

"I think there's a longer one in China."

"Why do foreigners give their animals names, like Tom and Jim?"

I laughed. He was worried about his trip the next day, so I explained what would happen at the immigration desk in Istanbul, and how the baggage carousel worked. Farah and Suleyman would be waiting for him outside.

Omar's father was sixty-three. Tomorrow's flight to Istanbul would be his first time in the sky. He told us about the first time he'd seen an airplane. It was during the king's reign, when he was ten years old. The plane was bright yellow, with a single propeller, and had landed on the road in front of the government hospital. Someone in the crowd said that it was a crop duster, sent from Canada to deal with the plague of locusts that year.

When Jamal got up, Omar and I stared at each other. "I've never seen your father friendly like this," I said.

"He's a stranger to me," Omar answered. "The other day, he said 'thank you.' I can't remember when he's ever thanked me for anything."

When we went outside, I saw Jamal pacing in the darkness between the fig trees. The summer air was so mild and dry you could barely feel it.

Omar went to open the gate. Jamal came up to me, the streetlight flooding across his worn features.

"You can't take him to Canada or America, can you? Isn't there some way?"

"It isn't possible. Maybe if he goes to Europe first."

"But what about the military? Could you ask them to consider him? So many others have gone that way."

"The American visa wasn't up to me. They rejected him because he didn't have the right documents."

"Do you want to help him?"

"I do."

"He was the child who first wanted to go. He was the one who most wanted to go. And now, he's the last one left."

"He'll find his way," I said, looking at Omar. "And I'll help him if I can."

The next day, I went with them to the airport. Omar had dressed his father in a pastel golf shirt and slacks. The Western garments made him look even taller. We shook hands, and then to my surprise Jamal embraced me.

"Take care of Omar," he whispered.

BY JULY, I HAD WAITED half a year for Omar. Soon I'd have to give up and go back to New York empty-handed, leaving him behind in Afghanistan. It was such a waste. The summer was slipping away. Maryam called from Istanbul weekly, hoping he'd changed his mind about leaving. They'd found an apartment there in a neighborhood with lots of Afghans. There were rumors that the border would open again.

Omar would still wake up early to follow Laila to school, but he didn't approach her. Just seeing her quieted something in him. He would park for an hour outside the gate, smoking, watching the students come and go. He told himself that she was trying to drive him away because she wanted to obey her father. But he was certain she still loved him.

He began to talk to me again about our plan, about going to Europe and then coming back for her, after he had something to show her father. Maybe he would go to Italy, where he heard Afghans got their residency papers the fastest. Germany or Sweden would take too long. People languished for years without getting asylum. He didn't have that much time. I told him I was ready.

One day he got a call from a number he didn't recognize. It was Laila.

She told him she was sorry for what she had said to him, outside the university. "People were watching."

"I'm sorry, too." It was his fault for making a scene. He told her he'd been thinking of leaving for Europe, while he still could. If he went, would she wait?

"Go," she told him. "Go, and come back for me."

Part II

The Road

6

Omar leaned against the escalator railing and looked down into the depths of the shopping mall. "Since we became thieves, the moon came out," he said glumly. The Persian expression meant something like *We missed the boat.* We'd just visited the travel agencies on the lower floor of the Gulbahar Center. Six months ago, when Europe's borders were still open, they had been packed with smugglers, some offering all-inclusive trips from Kabul to Germany, guaranteed or your money back. Now we'd been told that the farthest you could go from here was Istanbul.

Imagine the cities of the world connected by a network of paths that measure not physical distance but danger: the risk of getting arrested, stuck in transit, scammed, kidnapped, or killed. For the underground traveler, the shortest distance between two points is rarely a straight line; it might even be a flight halfway around the world to transfer in an airport with corrupt officials. The space between two people clasping hands through a fence could be wider than a desert.

For each line, there is a price. More money means less risk, and migrants take the shortest path they can afford. Only a fraction of the network is visible to any individual traveler, and the lines shift as borders tighten and people find new ways to cross them. The previous year, some migrants seeking asylum in Europe had discovered a loophole where you could bicycle through the Arctic from Russia to Norway.

For Afghans, the longest, cheapest route into Europe was across the land border with Iran, and then over the mountains to Turkey—an exceedingly dangerous journey. The short way was to fly to Istanbul with a visa, as Omar's parents had done. That was his plan, too, and since demand had dropped now that Europe's border was closed, he was able to haggle for a visa and flight at one of the agencies for $2,500, half of what his mother had paid a few months earlier.

The agent told him that it would take at least two weeks.

Once Omar got the visa, we would fly to Turkey together. From there, it would be up to him how he wanted to get smuggled into Europe. Most likely, we'd take a boat to the Greek islands. I was both nervous and relieved—I hadn't quite realized how badly I wanted this trip to happen. Now that Omar's house was empty, we spent our nights there playing cards and watching Turkish soap operas, dubbed and censored with blurred boxes that clung to women's chests like futuristic fashion accessories. When the news came on, the headlines that summer of 2016 were no longer about people in boats, but about the British voting to leave the European Union, about the Republican Party nominating Donald Trump as their candidate, and about a truck crushing people on the boardwalk in Nice. At the Democrats' convention, a man with his wife in a hijab beside him held up a copy of the Constitution, his voice breaking: *Go look at the graves of the brave patriots who died defending America.*

Sometimes two of Omar's friends from the neighborhood would join us. Zakaria was nineteen, a tall and rawboned basketballer. Born and raised as a refugee in Iran, he was from a Shia minority that had been historically oppressed by Afghanistan's Sunni rulers. Hazaras tended to have distinctly central Asian features, such as high cheekbones and almond eyes, and I was often taken to be one. Their faces marked them as Shia for murderous extremists like ISIS in their own country, and as Afghan refugees in Iran, where the police were constantly stopping young men like Zakaria. He got so fed up with his precarious status that he considered joining the Fatemiyoun, the Afghan militia that the Iranian government was recruiting to fight on the side of the Assad regime in Syria. The pay was good and you could win legal residency in Iran for your whole family. But he'd heard too many nightmarish stories about the fighting in Aleppo, and how the rebels would pull out your intestines with skewers if they caught you. Instead, he decided to flee to Europe, but was caught on the Turkish border by the Iranians and deported to Afghanistan, a land he'd never seen before. Now he was working cleanup on construction sites, getting stoned in the evenings and listening to Rihanna, dreaming of escape.

Omar's other friend Malik was a tailor who'd grown up in Kabul. He was twenty-one, slender with curly hair, and sat with his shoulders hunched forward, as if trying to take up less space. His father, a truck driver, had struggled to support his family and one day lost his mind. Now if the old man wasn't watched carefully, he would escape their house and vanish. During his fugues, Malik's father displayed a childlike trust in those around him and was soon relieved of phone and money. Omar and Malik would drive around searching for him, and usually found him at the taxi stand in Kot-e-Sangi, where the shopkeepers knew him from his driving days.

The previous fall, Malik had borrowed $1,400 from his relatives, just enough to pay a smuggler to take him to Istanbul the hard way, across the deserts of Nimroz and into Iran. It was a harrowing ordeal, he told us. He had walked without water for a whole day and, once he was in Iran, the smugglers had stuffed him into the trunk of a car. After two weeks he made it to the Turkish border, where the Iranian guards caught him and deported him back to Afghanistan. "I heard that it was dangerous, but I didn't think it could be that dangerous," he said. "I thought I was going to die."

Some part of me was curious to travel that desert road, but I knew Omar didn't want to take unnecessary risks. It would be dangerous enough getting into Europe from Turkey, now that the border had closed.

IT WAS A HOT AND sunny day, and Omar had the Corolla's windows down as we inched through traffic. In the blue sky above Kabul, the white surveillance blimp had a clear view for its array of cameras, and what was known as *wide-area persistent surveillance*, technologies like Gorgon Stare and ARGUS-IS. In Afghanistan, we were living in a fishbowl where the US government accumulated the trivia of our lives; for example, every cellphone call and message, each late-night sext, was collected into the MYSTIC program. Perhaps the very hairs on our heads were numbered. That day, the aerostat would have recorded our Corolla amid a sea of Corollas, slowing to walking speed and then to a crawl as we moved into the center of town. Along the riverbank, half-ruined and half-constructed buildings stood in a jumble of raw concrete and rusted rebar, their tinted glass smeared with kebab soot, the ground-floor shops arranged by genre: phones, gold, pots and pans. Omar parked at Mandawi Bazaar and we got out, entering a covered warren of one-room clothing shops, each with a merchant holding a phone in one hand, a fly swatter in the other. We were shopping for our migrant outfits.

In the Afghan countryside, people wore traditional tunics and pantaloons, and dressing otherwise could attract unwanted attention. Whereas in downtown Kabul, you saw as many men in jeans or suit jackets. There was a class aspect to it: some young professionals turned their noses up at *robe wearers* their age. But rich and poor alike knew that if you went to Europe, you had to dress like the people there.

Omar and I walked to the cheap part of the bazaar, in the open by the stench of the riverbank. The garments here came from factories in Bangladesh and Cambodia, wherever wages were lowest, and they were similar to what you found at bargain retailers like H&M and the Gap, which feed the West's insatiable demand for *fast fashion*. But though the style here followed trends in London and New York, it seemed amplified by distance: neon fleurs-de-lis on pastel checks, superlatives spelled with sparkles, and caterpillars of ornamental darning. I examined the tag on a pair of houndstooth cotton pants, printed on elegant, heavy-gauge cardboard:

BAROBRRY
Ever Since The Ending Of
The Epoch In Which The Grandee Toilette
Prevailed The Convenients
Succinct And Function Modem
Fashion
Tide Have Changed The Visage
Of The Men's Apparel And Their Life

Signifier had decoupled from signified; knockoffs became forms in themselves. There was a rack of CK belts: Cal Kreian and Calwine Klam. We bought floral shirts and distressed jeans, and then got back in the Corolla, and drove to Bush Bazaar, a sprawl of shops built from recycled shipping containers. These standardized steel boxes were one of the most important logistical innovations of our time; by greatly reducing the labor costs of global trade, they'd given an increasingly cosmopolitan character to consumption. Amazon was unthinkable without them. Bush Bazaar, named for the forty-third president, had sprung up soon after the invasion in 2001, specializing in objects that fell off military trucks, but it had since diversified into imported cosmetics, camping gear, and secondhand clothing whose fabric was often of better quality than the new stuff we'd seen in Mandawi.

In the 1960s, the anthropologist Louis Dupree was amused to see Afghans walking around in American surplus uniforms, complete with World War II campaign medals. Back then, 95 percent of Americans' clothing was made in the United States; today, 90 percent comes from abroad. As a result, many garments make a circular journey across the earth. When you donate used clothing to a local charity, it's typically sold in bulk back to the developing world.

We entered one of the container shops that specialized in used T-shirts and browsed through the universal opulence that was America. There were shirts for softball tournaments and family reunions, for towns where people partied on beaches, and towns where they enjoyed colonial history. There were shirts with jacked Christian angels, fast-food brands, a BARACK THE VOTE T-shirt, a National Rifle Association

T-shirt, and a shirt that said TELL YOUR BOOBS NOT TO STARE AT MY EYES. There were cuboid stacks of XXL and XXXL shirts in dimensions that did not obtain for any Afghan I'd ever known, except for one, or maybe two, warlords. "In America, the poor are fat and the rich are skinny," I told Omar, who shook his head in disbelief.

The white noise of English pervades the planet. I had journeyed unembedded into the heart of the Hindu Kush and taken tea with white-bearded elders wearing Playboy Bunny knit caps and Johnnie Walker sweatshirts. On my first visit in 2008, I spent an evening in a *chaikhana* in Ghor discussing whether dancing was a sin. My companion explained that, in his written works, Ayatollah Khomeini neither expressly prohibited nor permitted dancing. Halfway through our conversation, I noticed he was wearing a pullover with a caricature of a dandy brandishing a cane, above which it read, unbeknownst to him, MARLBOROUGH DANCE CENTER.

At Bush Bazaar, I bought a yellow Lucky Charms T-shirt. Omar picked a black one with a Harley under the Stars and Stripes, and LAND OF THE FREE underneath. Then I browsed through some Narwe Face gear and selected a thirty-liter backpack. On our way out, we stopped to gawk at a collection of knives in a glass case, everything from pocketknives with tweezers to spring-loaded, blackened switchblades.

"Should we get some?" Omar asked. I looked up from the cabinet. Pretty soon we'd be on our own in some rough places.

"You've been in knife fights before, right? How many people have you stabbed, anyway?"

"Many, I don't know. Ten people, or more." He started counting the scars on his arms and neck. "In Iran, in Pakistan, in Afghanistan . . . Forget it. I did it to defend myself, because I had to."

We left unarmed.

PRESTO! IN THE TWINKLING OF *an eye, so to say, I had become one of them. My frayed and out-at-elbows jacket was the badge and advertisement of my class, which was their class.*

At the beginning of his 1903 exposé *The People of the Abyss*, Jack

London outfits himself in secondhand clothes and sets out into the slums of London, among *a new and different race of people, short of stature, and of wretched or beer-sodden appearance.* The book was an immediate bestseller. *The young American writer has studied the people of the great East End by the same methods,* wrote one contemporary reviewer, *that an explorer might adopt in learning the characteristics of a savage tribe in Darkest Africa.*

Sometimes the lens we train on others shows us ourselves. Victorian journalists and social reformers called their undercover slum missions *going Harun al-Rashid,* after the caliph who dispenses justice incognito in *One Thousand and One Nights.* Sir Richard Francis Burton's racy translation of the Arabic folk tales was published in 1885, when he was already famous for his travels beyond the periphery of the British Empire. Most notoriously, the officer and virtuoso linguist had penetrated the holy city of Mecca—forbidden to non-Muslims— disguised as a pilgrim and carrying an Afghan passport, sketching pregnable Ottoman fortresses as he went. *Hating and despising Europeans,* he reported from Egypt, *they still long for European rule.*

Yet passing was not simply an imperialist vocation. Jamal al-Din al-Afghani, who advocated pan-Islamic resistance to the colonial powers, and whose mausoleum lies on the grounds of Kabul University, was really from Iran. When he presented himself as a turbaned Sunni cleric at the Ottoman court, he'd wanted to disguise his Shia origins. Afghans themselves wrote about passing as different sects or ethnicities, particularly in the nineties, during the time of the warlords, when being from the wrong group could get you murdered at checkpoints. *I watched them carefully,* recounted Ali Akbari in *The Illegal Journeys,* his memoir of emigrating to Europe. *I copied the way they washed their faces.* He tried to pass as Sunni, *but I made one or two mistakes. I had mixed up their way of praying with my family's way of praying.* The Hazara militiamen in Kabul had a shibboleth: a ball of dried yogurt, *quroot* in Dari, that they held up walking down the bus aisle. Many Pashto speakers couldn't pronounce the uvular plosive, *qaf.* They were taken off the bus.

I was working on my pronunciation. I'd been rehearsing my alter ego, a young Kabuli from a modest background, twenty-six instead of my own thirty-one, so I'd have less to explain. I taped myself and compared my voice to the tracks Omar had recorded for me:

My name is Habib. I'm from Kabul, from Shahr-e Nau. Well, if you know Kabul, I'm from Qala-e Fatullah. I lived with my mother, brothers, and sister in a rental house. The rent? Six thousand afghanis a month. I was working in a restaurant in Hajji Yaqub Square, but we didn't have a good situation. Like any other refugee, I left my country and came by the smuggler's road.

My task was aided by the fact that Afghanistan had so many languages and dialects, as well as people who'd returned from exile with foreign accents. For years now, blending in for my own safety had been a habit. In turn, passing as an Afghan had made me aware of how, as an Asian face abroad, I had to perform my own Western identity if I wanted to wield its privilege. The American ring to my voice, the way I made eye contact, the clothes I wore—these were levers that could shift the world. *Passing, in this sense, is a universal condition*, wrote Asad Haider.

I asked Afghan friends if they found my mimickry offensive, but none seemed bothered. "It depends on your intentions," a journalist told me. I think some even saw it as self-improvement if I learned to speak the language of Hafez, if I could sit cross-legged and break bread, if I understood the rituals of Islam—"*Afarinet*," they'd say. Good for you.

In any case, there was no other way for me to travel underground with Omar. If the authorities arrested us and discovered that I was a Westerner, we'd be separated. Our smugglers might kidnap me for ransom. Omar would be in danger, too. But those knives in Bush Bazaar had got me thinking about the limits of what I would be willing to do, so I set a rule: I would only lie or break the law to keep us safe, and if it didn't harm anyone else.

ONE NIGHT WHILE WE WERE waiting for Omar's Turkish visa, the TV had shown burning buildings, angry crowds in the streets, and a tank crushing its way through traffic like at a monster truck rally, except the cars had people in them. There'd been a failed coup attempt in Turkey. "That doesn't sound good," said Omar.

A few weeks later, he came over to see me, his face ashen. The broker had returned his passport and money. The Turkish embassy put all visa applications on hold after the putsch. Omar wasn't going to be able to fly to Istanbul. We would have to cross the desert after all.

7

The smuggler had said to meet him in Paghman, a valley just outside of Kabul that was popular with picnickers. Omar eased the Corolla over a rut as the steep valley came into view. He'd been silent for most of the drive, still upset over being denied a Turkish visa. He had wanted to fly there in safety, but it was already late August and we were running out of time before winter. We had to go with plan B: travel more than two thousand miles overland from Kabul to Istanbul. To do that, we needed the right smuggler.

There are few figures more reviled today than the smuggler. According to Western politicians, their greed is at the root of the migration crisis. The Italian prime minister, Matteo Renzi, called them *the slave traders of the twenty-first century*. Of course, the smugglers don't see it that way. *Honestly, the people who help asylum seekers the most are people-smugglers*, wrote Dawood Amiri, jailed for his role in a deadly boat disaster near Australian waters.

In Afghanistan, where people had been fleeing war for two generations, everyone had a relative or friend who could put you in touch with a *qachaqbar*. The smugglers weren't shy, either. After a mutual friend introduced us, Karim said to come to Paghman, where we found him supervising the construction of an outdoor bathing pool. A big, jowly man with curly hair and pinky rings, he invited us to join him for lunch. We sat together on a platform, listening to the sound of the river below. I was nervous that he might spot me for a foreigner, so I let Omar do most of the talking. Karim was a Pashtun from the area, but he had a partner down in Nimroz connected with the Baloch tribes who took people through the desert into Iran.

Without borders, there'd be no smugglers. When Omar was little, it was easy for Afghans to go back and forth to Iran without paying more than a token bribe to the guards at the crossing. But in the nineties, Iran tightened the border, and Afghans started making short trips on foot near the frontier city of Zahedan, paying smugglers around $150 to get to Tehran. Then Iran built a fifteen-foot wall manned by armed guards, and migrants were forced deeper into the desert through Pakistan, into remote areas controlled by the traffickers and insurgents. Now the longer, deadlier journey to Tehran cost upwards of $700. Yet more people had crossed the previous year than ever, and their passage was estimated to be worth over half a billion dollars. Iran's wall had helped concentrate the industry in the hands of gangs who could invest in equipment, bribes, and manpower—the entrepreneurs of violence.

Over a dish of fragrant mutton *karahi*, Karim explained how the smugglers in his area worked under the protection of a powerful ex-mujahideen commander, a politician who could intervene with the police in Kabul if necessary, and in return collected payments and other favors. He told us not to worry; he'd sent his own two sons that way the year before. Now they were living in Germany. "The road through Nimroz isn't so bad," he told us. "Just don't run if the police stop you. Don't run, or they'll shoot."

We thanked him and said we'd think it over. Back in Kabul, we met with a few more smugglers; Omar was desperate to find someone who would take care of us if we ran into trouble in the borderlands. Then he heard about Agha Sahib, an Afghan who lived in Iran, who'd sent one of Omar's cousins to Turkey the year before. They spoke over Viber, a free, encrypted calling app; the smuggler claimed he could pay off the Iranian guards at the official crossing in Nimroz, so that we wouldn't have to detour through the desert. He asked for $1,300 each to get us all the way to Istanbul, a fair price, which we had to deposit in escrow. Omar was overjoyed that we'd skip the Pakistani desert. It sounded a little too good to be true to me, but we didn't have any better options.

"I thought the border was shut tight there," I told Omar.

"There's always a smuggler's road."

ON THE DAY OF OUR departure, Omar took the Corolla to the used-car market in Kot-e-Sangi where, despite its battered condition, he got three thousand bucks. "It was a manly car," he told me afterward, looking forlorn. That was it for farewells—his family was already gone, and his beloved was locked behind her father's walls. For the second time in his life, Omar was about to become a refugee.

The bus to Nimroz left in the early hours of the morning. When evening came, I laid out the robe and pantaloons which I'd wear in the borderlands, and stuffed the jeans and shirt from Bush Bazaar into my backpack. I added some dried fruit and nuts, in case we got stranded in the desert, along with a tourniquet and gauze compresses, since the border guards were so trigger-happy. The risks of going undercover were much higher now. The smuggling gangs were notorious kidnappers and had links with the Taliban. If they found out I was a Westerner, I could be worth millions of dollars to them. I was even more worried about the Iranian government. Though the police could be brutal during arrests, Afghan migrants were usually deported without further punishment. But if the Iranians realized who I was, I would probably be accused of spying. Some Americans who were arrested hiking on the border with Iraq in 2009 spent a couple of years in prison before the sultan of Oman was able to negotiate their release. I had explained to Omar that if the Iranians discovered me, he had to claim we'd met on the road and deny knowing my real identity. Otherwise he'd share my fate.

But how would they know? Afghan migrants often traveled without papers—they might be taken by robbers or the police—and Omar and I were both leaving our passports behind in Kabul. I took out the cheap smartphone that I'd bought for the trip and went through it carefully, deleting my call records. I checked the time; Omar wasn't picking me up until midnight. After months of waiting, it hardly seemed real that we were leaving tonight. I showered, then stood in front of the mirror: The Iranians didn't have my fingerprints. I wasn't tattooed, I was circumcised. Skin is an opaque surface; language is what betrays us.

My Samsung rang; it was Omar.

"What's up?"

"Oh, not much. Is it a problem if some other people come with us?"

"What?"

"Is it OK if Malik and Elham come with us?"

Malik was the shy neighbor whose father had lost his mind. Elham was Omar's maternal cousin. He explained that he'd offered to loan them the money he'd made from selling the Corolla, so they could come to Turkey. They were both close to Omar, and trustworthy, but it seemed last-minute, to say the least. Suppressing my irritation, I reminded myself that it was Omar's decision, not mine.

"Do whatever you want," I said.

In the end, when Omar showed up in a taxi at midnight, it was just Malik in the back seat. I got in beside him and we took off. When I greeted him he tried to smile but only managed a terrified grimace. I wondered if my presence, as a foreigner, was comforting. Probably not. Or did he think I had some special way of getting us out of trouble? Of course Malik had already gone through the desert, and knew the trials that lay ahead.

Omar and Malik were silent as we raced through empty streets, leaving their city behind. I breathed evenly, trying to slow my pulse. The months of preparation and patience would be put to the test now. If you were ready for it, the danger could give you a clarity of focus that swept away all other emotion, leaving you with a kind of grace.

We were headed to a neighborhood on the western edge of the city called Company. It was the starting point for travel south along the national ring road, Highway 1, which curved down to Kandahar, and then back up to Herat. Company's main drag, its asphalt shredded by truck traffic, was lined with cheap hotels called *mosafarkhana*. Behind them were feedlots and slaughterhouses for livestock, a maze of mud walls by a ravine running with offal.

We paid the driver and got out, making our way past families guarding piles of luggage, and villagers counting sacks of grain and cans of cooking oil. You could spot the other Iran-bound migrants by their backpacks. The *mosafarkhana* were lit with strings of colored lights, beneath which vendors sat at their stalls heaped with nuts and cigarettes, or sauntered among the passengers with trays of wallets and prayer beads. Ticket touts cried the names along Highway 1: Wardak, Ghazni, Zabul, Kandahar, Helmand, Nimroz, Herat. Beggars worked the crowd, some proud and elderly, others haggard junkies: "I'm praying for you, O travelers! Blessings on your voyage, O Muslims!"

There were many bus companies that plied Highway 1, but we bought tickets from Ahmad Shah Abdali, encouraged by the relatively modern-looking coach parked out front, a Mercedes O404, a model introduced in 1991. When the gates opened at 1:00 a.m., we joined the crowd jostling into the bus yard. There were more than a dozen glossy 404s with their cabin lights on, bound for different destinations. But where was the bus to Nimroz? "Back of the lot," an employee told us.

There we found a row of decrepit Mercedes O303s, a model from 1974. Our bus had a MASHALLAH sticker obscuring much of the cracked windshield, and a biohazard decal by the door. Its side read, in German:

Children's Paradise
Dobler Travel

and gave an address in Lower Saxony.

The bus conductor, stubby with a balding fringe, searched our packs and then wrote our seat numbers on them with a marker. "What's he looking for?" joked Omar, who I noticed had brought an extra duffel. "Who takes drugs and guns to Kandahar?"

It was like carrying coal to Newcastle.

For safety reasons, the government didn't allow the buses to leave before three in the morning. We tried snoozing upright in our seats, as hawkers and beggars shuffled up and down the aisle invoking the powers of God and energy drinks, while more passengers straggled in, mostly young men with backpacks.

At last, the overhead lights went off. There was choral revving, then a concatenation of horns. Starting with the newer 404s, the buses departed en masse. As we rumbled down the strip, more coaches poured from the other companies' lots, forming into a single, thunderous convoy. The drivers jostled for position; they called this part *buzkashi*, after the national sport, where teams of horsemen fight for a headless goat carcass. Picking up speed on the bridge over the ravine, the bus pack cleared the edge of the power grid, and plunged into darkness.

RATHER THAN A SOLID MASS that filled its borders, the Afghan state resembled a medieval empire, its thin ribbons of control stretching along roads and valleys into anarchic terrain. On Highway 1, the government was mostly limited to the towns, especially at night; in the rural areas, the men with rifles waving you down could be anybody, police, bandits, insurgents, some combination thereof. At their flying checkpoints, the Taliban arrested and sometimes executed those suspected of working for the government; ISIS was killing people just for being Shia, that is, Hazara-looking. Highway 1 had many dangers. There were roadside bombs, of course, and the buses were notorious for reckless driving. The drivers modified the big German engines with an extra gear, so that they could hit eighty miles an hour down the two-lane highway, blaring arpeggios, swerving around blast craters and tanker trucks. The drivers did the fifteen-to-twenty-hour trip in one shift, often with the help of hashish or amphetamines.

At dawn, when we entered the rocky valleys of Ghazni Province, we passed two blackened bus carcasses, each with a stone pile adorned with flags. Seventy-three people had been killed here three months earlier, when the buses collided head-on with a fuel tanker. The warped chassis were too bulky to bother hauling for scrap. All the way to Herat, you passed these auto memento mori, some long rusted, others with heaps of glass not yet lost to the wind.

As daylight flooded the cabin, the passengers revived themselves with energy drinks and pinches of *naswar*, a caustic local tobacco inserted inside the lip. I cadged a Pine off Omar and listened to the banter in the aisle.

"You'll be amazed by Turkey," an older kid was saying. "What vices do you have?"

"I don't have any vices but hash," said another.

"It's hard to find hash there, but there's lots of drink. Don't say you won't try it until you've been tempted."

"Well, if it's in my hand, then I'll have no choice."

The day got hotter and the talk died out. The next province, Zabul, was desolate, a flat duotone of tan and brown. On the northern horizon we could see jagged peaks where the high valleys were ranged by nomads and their flocks in summer, the thin air scented with flowers which would lie beneath snow come fall, water that fed the plain in spring.

The villages came more frequently as we approached Kandahar toward midday. The driver opened the door for hawkers as we slowed; a child came toddling down the aisle with a tray of silver juice pouches, and leapt off when we slowed for a pothole a few miles later. Nearing the city, we passed open ground pitched with hundreds of tents: Kuchi nomads, come down from the mountains with sheep fattened for sale. A flatbed tractor-trailer sat out on the hardpack as a family and their livestock climbed aboard, the women's magenta robes washed out in the glare.

The name Kandahar was said to come from Sikandar, Alexander the Great, who founded a city here twenty-three centuries earlier. A hundred years later, Emperor Ashoka expanded his Buddhist kingdom until it stretched from Kandahar to Bangladesh. Archeologists discovered an edict of his nearby, written in Greek and Aramaic, then the local languages. The stone tablet recounted the emperor's conversion to pacifism after his conquest of Kalinga: *A hundred and fifty thousand persons were captured and deported, and a hundred thousand others were killed, and almost as many died otherwise. Thereafter, pity and compassion seized him, and he suffered grievously.*

In Kandahar City, rickshaws buzzed in the roundabouts. It must have been a hundred degrees, and the trees were limp with dust. We passed a billboard with an image of a boyish-looking man in a lieutenant general's uniform: Abdul Raziq, the border commander whom Omar and I had made our first trip together to report on. It was best that Raziq didn't find out we were here. He wasn't a fan of my work, although it hadn't affected his career much; he'd been promoted and put in charge of the entire province's police force, and he and his cronies had grown immensely wealthy. He was courted by elites in Kabul, but he'd also become a folk hero to those who hated the Taliban, and you could find his photo pasted in taxis and checkpoints across the country. After his promotion, the UN had reported that *officials in Kandahar province have increased the level of brutality and the use of torture and cruel, inhuman, or degrading treatment.* A couple of years back, a top American general had been photographed with his arm around Raziq, grins on both their faces.

It was lunchtime but despite the passengers' protests, the driver refused to stop. We kept driving west, and later that afternoon, crossed the Helmand River, where the road that forked south to the provincial capital was cut off by the Taliban. I saw another familiar face on a billboard: Commander Hekmatullah, the district police chief killed by a roadside bomb last year. Posters of heroes became posters of martyrs, and their ink faded pale blue in the sun. I had met Hekmatullah in 2014, during that season's record-breaking opium harvest, which had more than doubled since the year 2000. He'd kindly sent some of his men with me so that I could tour the edges of his district where the poppies started, fields that ran deep into arid wastes cultivated with bore wells and pumps, increasingly solar powered.

One hundred years earlier, after decades of panic over Asian immigration and opium dens, Congress passed its first drug prohibition law, the Harrison Narcotics Act of 1914. Since then, the war on drugs had forged causal chains that looped the globe. Under US pressure, first Iran and then Turkey banned opium cultivation; in the 1970s, Afghan farmers started to supply the world market, and production took off under the mujahideen. On my trip two years earlier, I stayed with a farmer in Marja in the old American-built canal zone so that I could watch the *nisht* harvest, when migrant laborers came from across the country to work the fields. Each olive-colored bulb was scored by hand in the evening, then scraped for sap in the morning. A good crop could be harvested thrice or more. Helmand was dry; in the 1950s, the United States funded two leviathan dams to help settle nomads along the river; now the deserts bloomed with a thousand Chinese panels. The industry consumed water, fertilizer, and people: sons arrested in Iran as drug mules and sentenced to hang, daughters mortgaged against bad harvests. But for life to be cheap, it must have a monetary value. My host had shown me some of his crop, a polyurethane bag of what looked like molasses. I hefted the basketball-size mass and inhaled its grassy scent, thought of its μ-opioid agonism. The Afghan farmer would sell it for a few hundred dollars to a local trader; by the time his harvest traveled the smuggler's road and was sold by the gram as heroin in Europe, it would be worth more than a hundred thousand.

IN NIMROZ PROVINCE, THE BUS left Highway 1 and took the road south toward Zaranj. We were now on the extreme verge of the Iranian Plateau. Beside us was the Dasht-e Margo, the Desert of Death, a flat expanse of basalt where summer temperatures could exceed 120 degrees. It was six thirty by the time we arrived in the provincial capital of Zaranj and the setting sun turned the dunes on its outskirts ochre. It had taken us fifteen and a half hours to travel nearly six hundred miles from Kabul. The bus stopped at a police checkpoint and then pulled into the depot. Hawkers gathered as the passengers descended on stiff legs.

Bastani, bastani! A boy was yelling the Iranian word for ice cream. Another held up a tangle of charms marked with Quranic inscriptions. Malik and I smirked but Omar bought one and hung it around his neck.

We walked onto the main avenue, past money changers' booths with tattered stacks of Iranian rials. The two-story buildings were made of unpainted cinder blocks, their neon signboards gaining auras

in the gloam. A west wind gusted down the avenue, picking up scraps
of plastic.

Now we had to find the safe house. Omar called the number that
Agha Sahib, our smuggler in Iran, had given him, and got directions.
We squeezed aboard a rickshaw—a motorcycle with a cab welded onto
its chassis—and set off with a jolt down the road, halting a couple of
miles later, just short of the town's main square.

"I stayed around here last time," Malik remarked.

Up the road, a man in a mauve robe waved to us, and when we
approached introduced himself as the manager of our safe house. He
was Pashtun and looked to be in his thirties. We followed him through
an open gate and down an alley, where four teenagers with backpacks
were already waiting. We entered a low doorway on the right, into the
safe house, a walled compound around a narrow yard. The main room,
about forty feet long, had wooden beams under a low mud-and-straw
ceiling; the far end was partitioned by a waist-high wall and served as
the kitchen, presided over by a tough-looking cook in a powder-blue
robe.

The manager bade us to set down our bags and make ourselves
comfortable. He said we were part of a group called Arya. We had to
remember that name. That was how the different gangs along the way
would know whose *mosafarin*, or travelers, we were, as they relayed us
along like a package, their fee per head paid by the originating smuggler, Agha Sahib, who would collect the money we'd put in escrow
once we made it to Turkey.

"Go and get whatever you need from the bazaar, water, food," the
manager told us. "The smugglers might call you to go in an hour, or
they might call in the morning."

We put down our bags in the corner. I glanced at the four teenagers; by their accents, they were from the east, and had the look of
villagers. "Go ask him which road we're taking," I said, gesturing to
the manager. Omar returned with a stricken look.

"He says we're going through Pakistan."

Omar took out his cell and called Agha Sahib in Iran.

"He says that the direct road is dangerous at the moment, so they're going to send us through Pakistan. It's safer that way," Omar told Malik and me. "What do we do?"

I looked at Omar, who'd turned pale, as if he were about to be sick. I thought that we should continue. We'd already risked our lives to get to Nimroz; the detour through the desert wasn't a deal-breaker for me. I'd half expected it, but clearly Omar hadn't. He'd believed the smuggler, which was wishful thinking on his part. I could tell he was thinking of backing out.

"What can we do?" I said. "We don't have any choice."

"If you're willing, so am I," said Malik.

"Look, we'll be fine," I said. "OK?"

Omar nodded reluctantly. We sat for a moment, dazed. Then I remembered that we needed to change some money and purchase supplies; I asked Omar to come with me, and we left Malik to watch our bags.

"I trusted Agha Sahib," Omar said, once we were on the avenue. "He said he would send us straight to Iran."

We continued toward the town center. Zaranj had boomed over the past decade, not only from the traffic in people and drugs but from legal imports of fuel and cement. An Afghan friend of mine had worked here in the nineties, when it was a flyspecked, three-street town; he told me the shopkeepers had never heard of garlic. Now the stores we passed were hung with backpacks and ski goggles, useful for sandstorms. Omar bought a pair of silver sneakers for Malik, who didn't own a decent pair of shoes, and then we changed a hundred dollars into Iranian rials, getting more than three million back. The kids at the booth recommended that we take half in half-million notes, which were easier to conceal in your waistband or glue inside the sole of your shoe.

Just beyond us to the west was the Iranian border, with the official crossing over the Bridge of Silk. Ever since the Iranians had built the wall, the smuggler's route ran south into Pakistan, where the *120-day wind* whipped the sand into an impenetrable murk all summer. The region was known as Balochistan, and the Baloch were a people divided by borders. Entangled in the drug and espionage wars of the tri-border area, they were alternatingly repressed and co-opted by the state. It was the Baloch who took travelers across the shifting desert sands.

At the safe house, more travelers had arrived. We sat for dinner at a long plastic tablecloth on the carpet as the cook's boy walked behind us, setting down strips of naan and bowls of chicken broth. Afterward, as Malik tried on the sneakers, I lifted the extra duffel that Omar brought. It felt like it was filled with stones.

"We should bring only one bag each," I said. "What's in here, anyway?"

Omar opened it. He'd brought the contents of his bathroom cabinet: a brush, a family-size bottle of shampoo, cologne, and various lotions.

"What did you bring this shit for?" I said. Malik laughed. "Hair spray?" I yanked the can out of the bag. "Shampoo?"

"We need shampoo, brother."

"Omar, no." We wrestled over the hairbrush, and then he snatched back one of the bottles of cologne.

"This is expensive."

"Really, you'll regret it. Bring as little as possible," Malik said, trying on his new sneakers. "You'll be walking for days."

After jettisoning the extra duffel, and most of Omar's beauty products, we redistributed our gear until our packs were of roughly equal weight.

Two men entered with bundles slung over their shoulders.

"Are you Sayed Ahmad's travelers?" the bigger of the two asked in Dari. He had a flaxen bowl cut, and a toothy grin.

"No, we're Arya."

"Well, anyhow, look after my bag, will you?"

After twenty minutes, he returned and sprawled next to us. He announced that it was his seventh time going illegally to Iran.

"Wow. How many times have you been deported?" asked Omar.

"Six times."

"Isn't it dangerous?"

He laughed. "It's the most dangerous road there is. You're playing games with your life, brother. You're in the hands of thieves and police. Human life has no value to them." In his northern drawl, he launched into an exposition of the perils ahead: car crashes, shootings, dying of thirst in the desert, and abuse by Pakistani police, who, he said, made even the Iranians seem merciful.

"The Pakistani police?" Omar cried. "The Pakistani police!" he repeated, gaping back and forth at Malik and me.

"The Pakistani police are waiting at the border. They'll search your pockets and your bag and take whatever they like."

"Whatever they like?"

"And you don't dare say anything. One word, and they'll beat you like a dog." The man stood, clasping my shoulder. "Brother, we're like a football kicked up the field. Whether we make it into the goal, that's our luck." He laughed again and went outside.

"Forget it, that guy likes to talk," I said.

"What about the Pakistani police?" Omar replied. "I didn't think we were going to Pakistan."

It wasn't a good idea to talk like this in the safe house.

"Let's have a smoke," I said.

Omar and I got up and went into the empty lot at the end of the alley. I lit our cigarettes, and we stood in silence for a moment. Atop the roof of the safe house, we could see the silhouettes of our smugglers, sipping tea and keeping a lookout.

I tried to think of what I could say to rally Omar. He'd once been up for almost anything, but as the years went on, I found him less eager to take risks on our assignments together. He'd had too many close calls and seen too many bodies. Of course it was different when you were doing it for pay instead of glory, but for me, repeated exposure had the opposite effect. Something had switched off, the *emotion-recording apparatus*, as Robert Graves called it. I understood that I was probably damaged, but at the time it seemed useful for working in places like Afghanistan and Syria. I thought I could separate feeling from reason. And it was natural, but not rational, for your fear to spike when it came time to act.

Omar spoke first. "The road is too dangerous, brother. I never wanted to go to Pakistan."

"We're in Nimroz. It's too late to turn back."

"It's not too late. We're still in Afghanistan. We can go back to Kabul."

"I'm not going back."

Malik emerged from the lot's deep shadow and joined us. "Listen, Omar, every danger that he talked about, I've seen. You're worried about the Pakistani police? When you cross the border, the police are waiting. They count the number of people in the truck, and they count the number of bills the smuggler puts in their hand. That's it. We passed through three police checkpoints and no one laid a finger on me." We could barely see the whites of his eyes. "Don't worry about the police or thieves. Worry about the truck flipping over. Worry about how dangerous the driving is, by God, you'll go on roads so high up, you'll look down and think, How did I get here?"

"Thank you, Malik," I said, interrupting. "He's right, Omar, that guy was talking *kos-e shir*. He's just showing off. Forget about him, OK?"

Omar nodded reluctantly. It was difficult to objectively measure risk in these situations, and even if you could, whether you were willing or not came down to gut decision. You tended to look at the next person's fear to see if it was more or less than your own. We had to act confident for each other's sake.

Back in the safe house, I realized we had forgotten to get water. I asked Malik to come with me to the bazaar. "Malik, you've been on this road before," I said, as we walked down the street. "Hundreds of people go every day. It's not that dangerous."

"It's ninety-nine percent dangerous," he answered. "But I'm willing if you are. We've come to Nimroz, and we can't turn back. What will we say to our families? We'll look like cowards."

"Listen, Omar's scared. How's he going to react the first time we're faced with danger? I need you to reassure him."

But when we got back I saw, to my dismay, that Omar was deep in conversation with the flaxen-haired man from Kunduz, who'd boasted of the horrors ahead. They were lying parallel on their sides, their hands propped up on their chins with their faces close together, like two boys at a sleepover. I couldn't see the stranger's face, but Omar's was rapt.

I sat down next to them.

"Have some naan, it's the best naan," the man said, turning and placing a big flake atop my backpack.

"No thanks, I've already eaten." He kept insisting, so I tore off a small piece, which glued itself to the roof of my mouth. I had to open a bottle of water to wash it down.

"So where are you from, my friend, what place, what people?" he asked.

"Kabul."

"Where in Kabul?"

"Shahr-e Nau. Where are you from?"

"Kunduz."

"From the city?"

"From the city. Have you been?"

"Yes." A scene from our last assignment there kept playing in my mind's eye, unbidden. Our photographer and I had gone on patrol one night with the Afghan special forces, along the stalemated front line. There was an exposed road that we had to dash across, but the Taliban snipers were active only in daytime, the soldiers said. I was jogging a few yards behind Vic when I saw the luminous tracer rounds bracket him at thigh level, felt them snap around me. The sniper was using a PKM, a light machine gun. A few seconds later, we were behind cover. "They got a night-vision scope, those *kos-e zan*," the team sergeant spat, apologetically. It had been a matter of inches, but we'd escaped unscathed.

"There's two places that I like best in Afghanistan," the stranger was saying. "Kabul and Mazar. Have you been to Mazar? You can stay out all night there, and no one bothers you." He smiled dreamily. In Kunduz City, the streets were empty after dark.

"What kind of education do you have?" he asked, after another pause. I stared at him. His face was tilted sideways and mashed up against his palm.

"Grade ten," I said. "You?"

He sighed. "No, I'm not educated." His haphazard questions were making me nervous, as if he could tell something was strange about me. Was I being paranoid? Shouldn't I be paranoid? The seed of fear had sprouted.

"I'm going to have a cigarette," Omar said, standing up. "Habib, can you come with me?"

From his expression, I knew he'd made up his mind.

"Look, from the very beginning, I said I didn't want to go through Pakistan. It's my fault for trusting that damn smuggler."

"Omar, we came here together to go to Iran."

"I'm not going to Pakistan. I'm not."

"You knew about all these dangers beforehand. Didn't you listen to Malik's story? You're just afraid now because the danger's in front of you."

"There's a difference between danger that comes upon you, and danger you choose to walk into. Why should we risk our lives? What good is Europe if you're dead?"

"It's the smuggler's road, of course it's dangerous! What else did you expect? You should have thought about it before we got on that bus."

"I can't risk my life, brother. Don't risk yours. It's not worth it."

"You're abandoning Malik and me. What kind of friend are you?"

He looked down at his feet. I felt an icy fury rising in me. We walked back to the safe house in silence. I blinked in the wan fluorescent light, and then asked Malik to come outside with me.

"Omar's going back to Kabul," I said.

Malik kicked viciously at a stone. "*Lanati!* He's the reason I came here. He asked me to come. What kind of man is he?" We stood, looking at the mud walls around us. "My mother begged me not to go. My brother begged me not to go. They were so upset. How can I face them now?"

I had one last card to play. I figured Omar wouldn't let us go on without him. "Malik, if you still want to go, then I'll come with you."

"I can't go without him. I don't have the money to get to Turkey. I only have enough for Iran."

"Forget about the money. I'll pay for it."

"But I came here because of Omar. I thought we'd go together. I thought that there'd be three of us, and then it wouldn't be so bad. With three, you can face what happens together."

"We'll be OK with two. You've done it before. It's not so dangerous."

He turned to me, his face sallow in the lamplight. He was trembling. "It's so dangerous that you can't even breathe." He shook his head. "I know what will happen to us. I'm afraid. I'm even more afraid than Omar is. But I have courage. He doesn't, damn him."

"Malik . . ."

He rose. "No. I'm going back to Kabul."

I watched him stalk off back to the safe house. The trip was a failure. Perhaps Omar would never leave Afghanistan. I'd spent nearly a year waiting for him. In the white glare of madness, I had the thought to continue alone. I stood for a while, coiling it inside me, until my anger turned to shame. I recalled Omar's question: Why should we risk our lives? I had far less of an answer than they did.

I went back inside. Omar and Malik were slumped in the corner. The other migrants were asleep, resting for the journey ahead. The teenagers were laid out side by side, their scarves shrouding their heads and torsos.

It was eleven. The bus back to Kabul left in four hours. Omar called our smuggler in Iran and told him that we were leaving. There was a serious family issue one of us had to deal with, Omar said, but we'd come back to Nimroz in two or three days. We slung our packs. At this, the stranger from Kunduz awoke and blinked at us. "Huh—what?" As we walked out the door, he bellowed: "Hey! Where are you going?"

"Kabul," I answered, glaring at him. He looked disappointed. Had he wished us well or ill?

In the narrow yard, we were stopped by the smugglers on the roof. "We'll be back in a couple of days," repeated Omar, squinting at the shadows above us.

"You don't mind if I make a call first, do you?" said one of them. It was the safe house manager. It was a rhetorical question, but he added, apologetically: "So they don't get mad at me for not telling them."

The burly cook was blocking the exit.

"Three from Arya," we heard the manager say. Then, after a moment, "OK." And to us: "How many are left?"

Malik looked back into the safe house and counted the bodies. "Eleven."

We pushed past the cook, who followed us into the alley: "Hey!" We turned. "Pay for your dinners."

We forked over some cash. Mollified, he accompanied us to the gate, advising us on where to buy our bus tickets. We filed out onto the road.

"Wait." He leaned out, his harsh features raked by the street lamp. "Listen to me. I'm going to tell you something because I'm Baloch. The road is bad now. Twenty people were killed on it the other day. Those guys just are sending people out to earn commissions. I'm not Afghan, I'm Baloch! They're Afghans, and they're doing it to their own people. Come back after Eid. The situation will be better then, God willing."

We took a rickshaw back to the depot and waited. The bus was only half full, and I took a seat by myself. We pulled out and headed north. Outside my window the Desert of Death was sunken in darkness. But up above a drone or satellite was looking down. Could it see those who'd lost their way in the wilderness?

Our bus would pass through the battlefields of the south again but I was too tired to care. Yet I couldn't sleep; pangs of remorse were starting to gnaw at me. Instead, I listened to the other passengers talking quietly in the aisle. Some had been deported from Iran, caught at the border or during immigration sweeps. Others had planned to cross like us but lost heart. One round-shouldered man, his beard streaked with white, said that he was a farmer from an area in Baghlan controlled by the Taliban, north of Kabul. His teenage son had wanted to find work in Iran, and when he wouldn't grant permission, the boy had run away with some friends to Nimroz, where they indentured themselves with a smuggler against their future wages, a common arrangement. But there was a dispute between gangs, and a rival had abducted the teenagers and locked them in a shack in the desert. When the father and the other boys' relatives found out, they'd banded together and came down to Zaranj, where they forced the first smuggler to sell his car and some land to ransom the boys. His fugitive son was safely in Iran, working, and the father could go home.

His story concluded, the man fell asleep with his forehead against the seatback; he awoke hours later in Zabul, with a mark on his brow like the devout have from prayer.

8

It was time for plan C. At the Kabul airport, Omar insisted on carrying my bags up to the checkpoint where two cops stood frisking passengers. We had a final smoke together, neither of us making eye contact. A week had passed since our failed trip to Nimroz. If Omar wasn't willing to cross the desert, and he couldn't fly to Turkey without a visa, then there was one option left: he could get an Iranian visa and go to Tehran legally, and then cross with smugglers into Turkey. He hadn't done this in the first place because he didn't want to trek through the mountains on his own. I couldn't fly with him to Iran—even if I was granted a visa, the government would assign me an official minder. But now Malik was willing to go, since Omar was lending him money. I would fly to Turkey on my own passport, and then wait for Omar and Malik near the border with Iran. Once they made it across, I would join them and we'd meet up with Maryam and the others in Istanbul. Of course, there was no guarantee that Omar and Malik wouldn't get caught and deported back to Afghanistan.

When it was time to say goodbye, I turned to Omar and saw on his face once more the fear of being left behind. We embraced each other.

"OK, dear brother, see you soon," he said, and grinned.

My bags on my shoulder, I joined the other passengers walking toward the terminal, a flood of tenderness and regret washing over me. I'd already apologized to Omar for how I'd acted in Nimroz. I was treating this trip like another assignment where I was in charge. But if I was going to follow Omar as a journalist, which was my justification for going undercover, then I had to let him make his own decisions. Yet I could hardly be objective when it came to my friend, especially when both of our lives were on the line. I had never felt so confused about my role.

Once we were airborne, I took a deep breath and leaned back in my seat. The calm of the cabin was soothing; I could organize my thoughts in the hours ahead. I turned on the screen in front of me, and toggled it to show an icon of our plane flying above the earth. The names of cities floated nearby. As a child, I'd been fascinated by globes and atlases, and my father had taught me how to navigate with a compass on sailing and camping trips. Our flight would cross the Caspian Sea near Baku, on a path that appeared curved onscreen but was actually a straight line over a sphere. Flat maps necessarily create distortion; on the Mercator projection in front of me, the tropics appeared tiny and North America and Europe huge. Although space seems stable, like time it can expand and contract relative to our motion. I traveled the twenty-two hundred miles from Kabul to Istanbul in six hours. The city grew in my porthole, a web of light that spanned the horizon, cleft by the Bosphorus, the navel of the world.

I had a two-hour stopover at Atatürk Airport, but I wasn't staying in Turkey that day. I had decided to drop off my laptop and other possessions, including my second passport, at a friend's house in Italy so that I could collect them once our trip was over. I'd stay there overnight, and then fly back to Turkey. It was hard to say how long I'd have to wait at the border for Omar and Malik, but I wanted to be ready.

When we disembarked in Istanbul, I followed the signs onto conveyor belts and escalators until they brought me to the check-in desk at the Turkish Airlines lounge, reserved for those traveling in first or business class, as well as economy passengers like myself who'd merited gold status through their rapacious air travel. The lounge had a two-story atrium with a floating staircase; I descended to the buffet and selected a freshly baked *pide*, then poured myself an Efes from the beer tap. I settled into a recliner and ate, watching a bank of screens which displayed news and sports from different time zones, the seconds ticking in unison. Around me, the staff whisked away our detritus, vanishing behind doors that concealed a series of work spaces connected by freight elevators and corridors, a second airport imbricated with the first.

Turkish Airlines served the most international destinations of any company, and Istanbul was one of the world's busiest hubs for air traffic. Out the window, I could see the white-tipped stabilizers of jets parked along the terminal, like a row of sails. When the philosopher Michel Serres looked out such windows, he saw modern messengers: *Angels of steel carrying angels of flesh sending angels of signals over angels of airwaves.*

I'd often flown through Istanbul, especially since I started covering the war in Syria. From here, it was a short domestic flight to Antakya or Gaziantep, where just across the border, cities were being reduced to rubble by artillery and aerial bombardment, and people were being slaughtered by the thousands. The violence had spilled over into Turkey; just two months ago, three men in explosive vests with assault rifles had attacked this airport, killing forty-five people and sending passengers fleeing through the terminals and onto the tarmac.

When it was time to catch my flight to Italy, I entered the aortic flow from six continents moving through the corridors, the passengers in stilettos and sandals, board shorts and Umrah robes, beards that might have been jam-band or Taliban, hipster or Hezbollah. Turkish Airlines didn't fly to Trieste but they did go to nearby Venice. I didn't need a visa to enter the European Union, and at Marco Polo Airport the cop looked at his monitor and then stamped my passport without any questions, no doubt because there were no flags in the system. Increasingly, border authorities rely on automated profiling such as risk assessments based on advance passenger data, in order to allow trusted travelers through while the rest are examined more closely. *The speed of some is logically related to the slowness of others*, explained Tim Cresswell.

I took the train to Trieste and spent the night at my friend's house. The cheapest flight back to Turkey the next day was from Ljubljana, a short bus ride away. My ultimate destination was Van, a city near the Iranian border. When we landed in Istanbul, I had to transfer again, but unlike the day before, I needed to clear immigration to reach the domestic terminal. It was still early when I landed, and the line was empty. I was halfway through the rope maze when a heavyset man hurried toward me and flashed his police ID. He took my passport, frowned at it, and then told me to follow him.

I'd never written about Turkey, so I wondered if this had something to do with my reporting from Syria. I knew the situation was still tense here after the attempted coup five weeks earlier, which had impeded Omar's visa. Several hundred people were killed and the plotters had attacked the hotel where the president, Recep Tayyip Erdoğan, was staying, but he'd escaped; afterward he called the events *a gift from God.* In the months to follow, fifty thousand people would be arrested in a purge by his government, some tortured in custody.

The cop brought me into a cramped interrogation room, where there were two other men in plainclothes, a skinny kid in jeans with a long, scraggly beard; and the other, older and seated, stocky and pot-bellied. He was clearly the boss, but since I didn't speak Turkish, the kid translated in English as best he could, which was not well.

They had me unpack my bag while they asked questions about what I was doing in Turkey. When I told them I was going to Van, which was in a Kurdish area, they got excited.

"But Van is very dangerous," the kid said. "Why Van?"

I explained I was writing a book about Afghan migrants, and that I also wanted to do a little tourism, see Lake Van and Mount Ararat, which was true, since I might be waiting a while for Omar and Malik.

The older guy was smoking a cigarette as he flipped through my passport, his frown deepening with each page. Each time I protested my innocence, he put a thick finger to his lips to shush me.

"What's your religion?" he asked.

"My family is Christian."

"Why only Muslim countries?" he said, holding up my well-stamped passport. I stared at him, stifling a laugh. Did they think I was here to join ISIS? They hadn't asked me about my reporting. Had they just seen a Middle Eastern–looking, military-aged male on the cameras, and decided to pull him in?

"I'm not just visiting Muslim countries, and besides, I'm a journalist. I'm visiting these countries for work."

"Where is your journalist ID?"

They had me there. I hadn't brought it, or any other paraphernalia. The kid told me to unlock the clean Samsung, which I did, and he took a quick look.

"You deleted all your contacts!" he shouted. "You're a liar. You won't tell us the real reason you're visiting. You will be deported from Turkey!"

I pleaded with them to google me, but the kid wouldn't translate and his boss started getting agitated, bellowing for me to get out, swinging his palm around like he was going to smack me. Stunned, I followed the heavyset cop outside. The whole thing had taken less than ten minutes.

He took me to the airport's deportation area, where there was a desk staffed by security guards. They told us the flight back to Ljubljana didn't leave until the next morning. The cop snapped on gloves and searched me and confiscated my cellphone and passport. He wanted me to sign some paperwork in Turkish but when I refused, he shrugged and led me into the cell.

It had a low drop ceiling and about thirty reclining chairs covered in paper liners, and looked like a dentist's waiting room, albeit one with a door that locked from the outside. A hallway in the back led to a bathroom and shower. There was a no-smoking sign, but around half the occupants were smoking.

I looked at the pay phone near the door, which had a list of embassy phone numbers next to it. Maybe I could still fix this somehow. "You can get a phone card from the desk," said an Arab man in a robe, who'd been watching me. I knocked until a guard opened, gave him twenty lira, got a scratch card in return, and tried several times to dial a friend in Kabul who was well connected in diplomatic circles. But I couldn't get the card, which required dialing into a complicated call menu, to work. Seeing my frustration, the man kindly offered me his own cellphone. I reached my friend and got a couple of contacts in Istanbul, but the weeklong Turkish holiday for Eid was approaching and people had already left their offices. I tried the Canadian consul in Ankara, who was apologetic and just as helpful as the phone card. Finally, I gave up and collapsed in one of the recliners.

"It's good to claim a seat before it gets busy in the afternoon," said the man, who sat down and introduced himself as Abu Haroun. He was slight, with the shaved mustache and untrimmed beard of a conservative Muslim. He spoke fluent, somewhat American-sounding English, but said he was originally from Yemen and had grown up in Saudi Arabia. He'd been living in Istanbul for years with his wife and children, but six months earlier, after he'd gone on pilgrimage to Saudi Arabia, the Turks, for some reason, wouldn't let him back in. The Saudis wouldn't take him, either. Yemen was shattered by civil war, so the Turks couldn't deport him there. So he'd been stuck here since then. No exit.

"You've been in this room for six months?"

"My lawyer thinks I will be out very soon, inshallah," he said, with a faint smile. There had been a dozen Yemenis trapped there with visa issues, but he was the last. He was the dean of the deportation cell. The others were just waiting for the next flight back to wherever they'd come from. Some were here in transit after being deported from somewhere else. There were only eight of us, but the number increased over the course of the day, as Abu Haroun predicted. Most people were from the Middle East or Africa, though there was a Chinese guy, and a young Russian who'd been denied entry to Ukraine. When one Nigerian man showed up, he came around and shook hands with everyone in the room. He told me he'd been denied entry by US immigration in Chicago, and was on his way back to Lagos, which he'd left two days prior.

Besides me, no one seemed particularly outraged to be there. My own anger soon turned into despair. Surely the trip was over now. We passed the day smoking and staring at the drop ceiling. I thought longingly of the Efes on tap in the business-class lounge above us. The women and children's cell adjoined ours, and the men would converse with their wives by shouting through the drywall. Sometimes the pay phone rang, usually someone's relative. I was summoned to take a call because I spoke French. The woman on the line wanted someone who wasn't there; he's gone, I told her. "But have they sent him to France or Cameroon?" No one knew.

I had a paperback with me, Toni Morrison's *Beloved*; when I felt calm enough to read, the time passed faster, my own misery fading. At night we snored in our chairs; at dawn, in the gloom of the exit sign, a security guard roused me for my flight. My belongings were returned, apart from my passport, which the guard handed over in a plastic bag to the plane's crew. After the other passengers boarded, I was marched to the back of the plane. When we landed in Ljubljana, a Slovenian cop was waiting for me on the jet bridge.

"What happened?" he asked, when we'd sat down in his office.

"I don't think Turkey is welcoming foreign journalists right now," I said.

He nodded sympathetically, and stamped me into the EU.

BACK IN TRIESTE, I NURSED my third drink at the Caffè San Marco and contemplated my predicament. Omar and Malik had gotten their Iranian visas and were already in Tehran. Soon, they'd attempt to cross with their smugglers and reach Istanbul. But now I was banned from Turkey. Clearing my name could take months, if it was at all possible. We didn't have that much time before winter set in.

For the first time in my life, I had an inkling of what the border meant to so many others: a wall between you and someone you loved. Sitting alone in the cafe, I asked myself what I ought to do. I knew I could just wait there in Europe for Omar, and then reconstruct his voyage for this book—assuming he made it. I knew I had pushed things too far in Nimroz, and that I was losing my professional detachment. I knew that I was gambling with my life and freedom in a way that might have seemed senseless to some people. But I couldn't abandon Omar. I had to get to Turkey. I needed a plan D.

Earlier that year, when the borders shut in the Balkans, tens of thousands of refugees were trapped in Greece and Bulgaria, where they were forced to live in filthy, violent camps. I'd heard that some had actually paid smugglers to take them back to Turkey. I could pose as an Afghan who wanted to be reunited with his family in Istanbul.

But when I thought about it, I decided I could do a better job on my own. Afghan migrants called it going *chocolati*, since crossing for free was sweet like candy. With Google Maps and a bit of luck, some refugees passed through Europe without smugglers, especially when it was just a matter of dodging an official crossing. But to travel over remote, trackless terrain, in the face of patrols and fences, you needed to navigate with a compass or GPS.

I pored over satellite and topographical maps on my laptop, searching for a way out through the soft underbelly of Fortress Europe. Since the EU had bottled up the Greek islands, many migrants were instead crossing overland into Bulgaria. With EU support, the Bulgarians were building a massive border fence, topped with concertina wire. But, from what I could discover online, the improved fence hadn't yet been extended to the Black Sea, where the border straddled the Strandzha, a rugged, sparsely inhabited nature reserve. I decided to cross there.

The next day, I walked into the gloomy arcade of Trieste's bus station, passing Cyrillic signs that advertised cheap fares home to the city's domestic and manual workers, and bought a ticket to Sofia, the Bulgarian capital. We rolled out through the piazza and headed away from the harbor, past the high stone buildings, toward the green hills inland. Until the First World War, Trieste had been the main port of the Austro-Hungarian Empire, on the north shore of the Adriatic. After he was posted here as the British consul, Burton, the explorer and spy, spent his final years translating *One Thousand and One Nights* and whenever he and his wife, Lady Isabel, were feeling *seedy or out of sorts*, they used to take this road up into the Karst Plateau where the Slavic hinterland began. Back then, you could travel all the way to Transylvania without leaving the polyglot empire, until the great war, when millions died for new nations, their byproduct the modern refugee: a fugitive who lacks, in Hannah Arendt's famous phrase, *the right to have rights*.

Today the earth is divided into nation-states. Some philosophers have argued that nations must exercise realpolitik, because they coexist in a fundamentally violent anarchy; others dreamed of a commonwealth of democracies where war becomes obsolete. *It can be shown that this idea of federalism,* Immanuel Kant wrote in 1795, from his ravaged continent, *extending gradually to encompass all states and thus leading to perpetual peace, is practicable and has objective reality.*

Kant's vision of peace has come closest to being realized by the EU, in a form of cosmopolitanism founded on free trade. Born in 1951 as a common market to unify the coal and steel industries of France and Germany—to make war *materially impossible*—it has since expanded from six to twenty-seven nations. In 2012, the EU was awarded the Nobel Peace Prize. *Over the past sixty years, the European project has shown that it is possible for peoples and nations to come together across borders,* the president of the European Commission said at the ceremony. *That it is possible to overcome the differences between "them" and "us."*

Heading east, we drove into Slovenia, where the massifs were moss-green against the fading violet sky. I am going on a holiday, I told myself. My reflection appeared in the window as the world beyond it dimmed to forms: streetlights, overpasses, exit ramps. At night, every bus trip looks the same. The orange-lit lots we passed might have been covered in snow somewhere in Ontario, as I headed home from college at Christmas back to the suburbs I treated as a prelude to the real story to come. *I would like to live my life as in a novel,* Gérard de Nerval wrote in *Voyage to the Orient.* The glass felt cold on my brow.

We'd gone from Italy to Slovenia without stopping, the change in the road signs' language all there was to show we'd crossed an international border, but when we reached Croatia, we came to a halt. The passengers got down to get their passports stamped. We were leaving the Schengen zone, the borderless core of Europe. Of the EU's many treaties, refugees will be made familiar with at least two of them: Schengen and Dublin, signed four days apart in the summer of 1990, one in a wine-making town in the tax haven of Luxembourg; the other in the capital of Ireland, which once lost half its population to famine and emigration. Between them the two treaties reflect globalization's central tension, which is that while capital and goods move freely, most people do not. Schengen abolishes border controls between its member states; trucks, bank transfers, and residents circulate unimpeded. Dublin reestablishes borders within Schengen for a specific kind of person, by mandating that asylum seekers apply in the first EU country they reach, and reside there until their case is decided. Boat people who land in Italy, for example, aren't allowed to move to France or Austria, and if they're caught, they can be deported back. In effect, Dublin creates a buffer zone for the richest members of the EU; Germany is surrounded by other Schengen states. The refugee burden falls heaviest on the countries lining the Mediterranean and Balkans, which form the carapace of Fortress Europe. Dublin is the price they pay for the wealth of Schengen.

After another five hours, floodlights again glared through our windows. We were at the Serbian frontier. Croatia had joined the EU in 2013 but was still waiting to be admitted to Schengen; Serbia was on the outside of both but was an eager candidate. Serbians had already been granted visa-free visits to the Schengen in return for adopting stricter migration and border policies, an example of what is known as the *externalization* of the EU's frontiers. Dublin precedes Schengen. Joining the EU means building walls against those still outside. Prior to entering Schengen, Poland had to end its visa-free policy for Ukrainians, as did Spain for the Moroccan workers who'd visited seasonally. As a result, the first migrant boats started crossing the Mediterranean in the nineties, short voyages across the Strait of Gibraltar by the *harraga*, those who burn—the road, the water, and their papers. Spain had strengthened its defenses, and now those wishing to enter the EU from Africa made longer journeys from Libya and Egypt to Italy and Malta. The EU's most important border was the Mediterranean; the sea was a moat that separated two continents whose vast gap in wealth was growing. The *border-industrial complex* supplied ships and surveillance drones while the EU's diplomats pushed migration treaties and security projects deep into Africa and Asia. Europe's so-called refugee crisis was better understood as a crisis of this border system, caused by the revolutions of the Arab Spring, which had toppled the gatekeepers along the Mediterranean's southern shore: Ben Ali, Hosni Mubarak, and, most important, Muammar al-Qaddafi, who'd once stood next to the Italian prime minister in Rome and warned: *Tomorrow Europe might no longer be European, and even black, as there are millions who want to come in.*

At dawn, our bus climbed through the mountains into Bulgaria, where we passed through immigration one last time, reentering the EU. In Sofia, I changed some euros into levs; one of the union's newest members, Bulgaria hadn't been allowed into either the single currency zone or Schengen. I boarded a train that went down the Maritsa River valley, where the plain widened into rolling farmland, passing villages with red belfries. Summer was gone, the wheat golden; at the end of the day, the rays behind us set the fields afire.

I'd traveled some seven hundred miles across the Balkan peninsula, from the shores of the Adriatic to the Black Sea port of Burgas. The next morning, I took a commuter van down the coast to Sinemorets, a resort town near the Turkish border. I got down and walked past boarded-up ice-cream stands to the hotel I'd found online. The receptionist frowned: an off-season single male, a sex tourist perhaps. As she checked me in, I asked about the bikes listed on their website.

"We have several bicycles for use by our guests," she said. "But be careful to carry your passport when you go out, especially in the border areas. You must know about the problems we are having with foreigners coming to Bulgaria."

I assured her that I did, took my key, and found my room. I put the backpack from Bush Bazaar on the luggage stand, signed into the wireless network, and checked WhatsApp. No messages.

Where are you? I wrote Omar.

THE NEXT DAY, I AWOKE to his reply: *We're leaving for the border tomorrow.* It was time to begin my reconnaissance. I pedaled through town and headed south.

Sinemorets was at the southern tip of an arc of beach towns along the Black Sea. Once known as the Red Riviera, the region had been a playground for the communist elite. Tourists came from both sides of the Iron Curtain, but some of the visitors, in particular East Germans, arrived with the secret intention of escaping to the West via Turkey.

In those days, border walls were associated with totalitarianism in Europe. Such physical barriers had been rare, historically, since they were expensive and couldn't deter an invading army or determined smugglers. What they were good for was keeping the masses in place. For the West during the Cold War, the Wall was the line between those who were free and those who weren't, and the people who crossed were refugees.

I cycled through a mix of fields and woods, with a few barns in the distance, as the terrain got hillier. I arrived at the border zone, 160 miles long and a mile or two deep, marked by a road that ran parallel to the Turkish border. During the communist period, the zone was forbidden to those without special permits and defended by fences and minefields. Researchers believe that more than a hundred people trying to escape to the West were shot and killed here. But border walls need surveillance to be effective, and in Bulgaria, the guards were assisted by a network of informants in the villages. In the resorts, a multinational operation surveilled the beachgoers; in 1964, the East German Stasi established a seasonal deployment, the Operating Group to Fight Human Trafficking.

In 1989, the Wall came down, and Eastern Europe transitioned to capitalism. Bulgaria removed the fences and minefields. *The theory then was that it was antidemocratic to have these kind of devices along the border,* said a former minister. And yet, as if the Wall's shattered concrete was sown like dragon's teeth, since then border barriers have proliferated. In 1989, there were only fifteen borders that were fortified with walls or fences in the world; by 2016, there were almost seventy, with more planned or under construction. Especially since the attacks of 9/11, these walls have been built in the name of security, and yet in practice they trace the line between rich and poor.

After Bulgaria joined the EU in 2007, the border zone was fenced again, this time to keep people out. The same guards who'd hunted their own citizens were now hunting migrants, who were routinely beaten and robbed and forced back into Turkey. Later that fall, not far from Sinemorets, an unarmed, nineteen-year-old Afghan named Ziaullah Wafa would be shot and killed by the border force.

I slowed down as I approached the checkpoint, where a border guard stood inspecting vehicles leaving the zone. I could see the barbed-wire fence running perpendicular to the road, through the fields and into the woods. Neither the guard nor the soldier sitting with a Kalashnikov stirred as I pedaled past, dressed in shorts and a floral shirt. I kept riding south, another two miles until I reached the town of Rezovo, named for the river which forms the border through the Strandzha Mountains. The nearest entry point to Turkey was thirty miles inland, but there was a spot at the mouth of the river where you could stare at the giant Turkish flag flying on the other bank. There were two middle-aged couples standing there, taking pictures. They were Bulgarian and looked old enough to remember the Iron Curtain; back then, wrote Kapka Kassabova, the border was *a near-mythical entity that couldn't be approached or seen.*

I SPENT THE NEXT FEW days cycling and hiking forest paths near the border, returning each afternoon to fling myself into the Black Sea, whose brackish waters tickled the briar scratches on my legs. In the evenings, I ate dinner at a barbecue joint in town; around nine, an armored car would pull up and disgorge several men in riot gear from the Gendarmerie, part of the extra security forces deployed on the border. Once, one did a double take at me and my sea bass, wondering, perhaps, if he was seeing Afghan migrants everywhere after too much time in the woods.

I was waiting to hear from Omar and Malik that they had made it safely to Istanbul. But Omar hadn't been answering my messages; I called Maryam, but she didn't have any news, either. Omar and Malik were somewhere in the mountains with their smugglers. I didn't want to delay any longer: a storm front was coming to the coast in a couple of days. I confirmed Maryam's address in Istanbul, and said that I would be arriving soon, though I didn't go into details.

"I might just show up at your door," I told her.

"May God be your companion," she said. "I'm praying for you."

I checked my gear: three liters of water, some euros and Turkish lira, a knife, a compass, and my Samsung, which had maps of the area I'd downloaded, and, like most smartphones, a built-in GPS. I discarded my swim trunks and went to bed early, feeling the calm that came after the decision.

In the morning, I checked out of the hotel and took a taxi to Silistar, a popular beach just north of the border zone. As I settled into the back seat, a familiar song came on the radio. I couldn't believe my ears. It was one of Omar's favorites: "Hero" by Enrique Iglesias. The platinum single starts off slow, just Enrique's croak accompanied by a guitar, but by the final chorus the crash cymbals, violins, and piano are rising as he cries:

> *I can be your hero baby, I can kiss away the pain.*
> *I will stand by you forever,*
> *You can take my breath away.*

There was a crunch as the taxi pulled into the gravel lot. I shouldered my pack and walked to the beach, past early birds on blankets, and climbed the rocks at the southern end. I had reconnoitered this route earlier; there was a juniper-scented path that ran along the cliffs. After half an hour, I reached a promontory where the border-zone fence ran to the sea. I walked into the woods and changed into jeans and hiking shoes, and then continued deeper into the underbrush where I found my landmark, a hollow with a crushed suitcase and a pair of baby sneakers.

The border zone began here. I crept through the bushes to the edge of the two-lane dirt road. On my side, there was rusty chain-link, while the other side had a newer, twelve-foot fence topped with barbed wire. On the hump between the lanes was a third fence, with horizontal strands, like a cattle barrier. Every quarter mile along the road, there was a telephone pole topped by a camera. At the base of this pole, where the camera had a blind spot, someone—a smuggler, no doubt—had cut holes in the fence.

Shimmying through the opening on my side, I checked for traffic, and then crawled across the road, under the cattle wire and out through the hole in the other fence, into the dark and quiet of the forest. During the communist period, some escapees mistook the start of the border zone for the actual frontier, and became easy targets for the guards on the other side. Of course, I knew I had to get all the way across the zone to the Rezovo River. I checked my phone's GPS and headed inland, across the highway to a wide hayfield where, off in the distance, a horse was watching me. I found the dirt road through the woods that I'd seen on the satellite map, a thin ribbon of brown that led all the way to the river. The tall strands of grass growing in the road seemed undisturbed, but I nearly stepped into a big pile of dung. Worried there might be mounted patrols, I decided to bushwhack, and set off on my compass bearing into the forest.

The bushes were full of lusty brown ticks, and clouds of mosquitoes and gnats wreathed my head. In defense, I doused myself in insect repellent and donned a broad-brimmed, net-veiled hat, in camo pattern, that I had found at a tackle shop in Sinemorets. So garbed, I wondered how I'd appear, a solitary figure headed south. A lost tourist? A Bulgarian vigilante? A smuggler going to pick up his clients by the river?

The rolling woodland was cut with steep ravines. Hidden underfoot among the fallen oak leaves were candy wrappers and plastic bottles. In one clearing, I passed a broad column of flattened grass that looked like the result of a flash flood, but I saw that it ran up and downhill, heedless of gravity. People must have been walking several abreast, a river of migrants.

After a couple of hours, the terrain began descending. I could hear what sounded like heavy machinery off to the southeast, which made me nervous. Now I smelled the earthy breath of the river as I followed the slope down, past more flattened bottles, Turkish brands, until I reached a road that ran beside the Rezovo River. Through gaps in the bushes, I could see the emerald water.

Walking out onto the road, I had a nasty shock: twenty yards away, half hidden in the bushes, was a gray-green border police tent. I leapt into the thornbushes by the bank and crouched, listening. Nothing. I hadn't seen any cops. But here I was easily visible from the road. I had to cross the river fast.

This was the point of no return. I'd been deported from Turkey as a threat to national security. Better to be an undocumented Afghan, whom the police usually let go. I pulled out a bottle of Bulgarian *mastika* from my pack, took a deep swig, and gagged—it tasted like aniseed-flavored gasoline. I took out my passport, then poured the liquor into a puddle and tried to ignite it, but it kept going out. Flustered, I started hacking out the pages with my knife, and lit them directly, building a pile. I stared as the flames blackened the colorful stamps and foil-trimmed visas—the past few years of work as a journalist. But I'd added too many leaves at once: they were smoldering, especially the thick covers, and a white plume hung above the road. I snatched the pages out, and then fed them back with singed fingers, one at a time, so that they burned clean.

My head was spinning from the smoke and *mastika*. I took off my jeans and shirt and stuffed them along with my pack into a garbage bag, which I blew air into and tied shut. Holding the bag, I waddled in my underwear through the thorns, and steadied myself at the river's edge. Lifting the bag overhead, I hurled it as far as I could, and jumped in after.

The water felt so cool that I put my head under a moment where it was dark. When I surfaced, I reached for the bobbing sack and pulled it along in a sidestroke. The Rezovo was only thirty yards wide here, and as I reached the placid midstream, a lakelike hush filled my ears. I could taste the river's iron; the water was green like the trees all around.

The Turkish bank was higher. I balanced on a slimy trunk and pitched the bag upward, scrambling after it through the mud, where I reached a dark grove. I put my clothes back on. When I looked back to the river, all I could see was blank forest.

The light was changing color. I checked my watch; it was five o'clock. My plan was to hide out until moonrise, and then walk to a beach town, Igneada, where a bus left for Istanbul in the morning. I found a spot behind a fallen tree and lay down.

My breath and heartbeat returned to baseline. The sounds of the forest came back in layers of crickets and birdsong. A squirrel twitched down a trunk and scolded me. I heard the beat of a partridge overhead, a loon's call, and what sounded like a deranged fox. The patches of sky dimmed between the leaves and grew indistinct. My limbs felt like they were sinking into the earth. When I closed my eyes, I could see the flames consuming my passport.

I started; something had rustled near my ear. The squirrel, I thought, but then heard an unmistakable *meeow*. I shot upright, gaping blindly. A cat, out here? It was so dark now that I couldn't see my hands. I fumbled for a stick I'd found earlier, and set it across my chest. I must have been dreaming. Or was I losing my mind?

The darkness was like a heavy quilt. The animals lay under it, too. As long as I stayed in the woods, no one would find me. Time collapsed to a single point. My last visit to Bulgaria had been eight years ago, when I got stranded hitchhiking and had to lie down in a field for the night, waking wet with dew. I learned to sleep rough in those days. When I arrived in Afghanistan later that fall, I'd gone to visit the ancient city of Balkh, whence the mystic Rumi came. The night before, I'd smoked too much hash with the boys at the Aamo Hotel and retreated to my room, panic-stricken. But when morning came, I got on a commuter van in Mazar, dressed in my new robe, and the other passengers looked right through me.

Balkh had been known to the Greeks as Bactra and rivaled Babylon in her day. Alexander the Great married a local princess, Roxana, on his way to India. At the beginning of the thirteenth century, Rumi left ahead of Genghis Khan's army and migrated to Anatolia, where the Sufi master taught *fana*, annihilation, a state where the ego and transient world were swept away, leaving only divine essence. *Turn back from existence toward nonexistence,* he wrote, *if you seek and belong to the Lord.*

I had gotten out at the village near the archeological site and walked onto the same plateau where the prophet Zarathustra had preached. The city was obliterated by time; all that was left was a vast ring of raised earth where the walls had been.

He spurns all that is perceived by the senses, and makes his throne upon the invisible.

Afterward, drowsy with heat, I found a park where some laborers were napping. I wanted to disappear; I lay down on the grass and fell asleep, the sun on my throat.

His love is manifest and his beloved is hidden.

In the creation myth, knowledge is exile. You learn you can't go back the way you came. When I saw poverty and war for the first time on that trip, I was struck with grief for the suffering of others, but only later did I see that I pitied myself, too, for living in such a world.

The face that you loved; why did you tire, when youth fled?

When the world's brokenness feels like our own, we search for something to mend it. But I'd lost my faith and felt unmoored from history; I thought that being present with people was the only solidarity I could offer.

If the beloved could be perceived, everyone would know love.

Now I was alone in a wilderness. This trip with Omar had gotten so screwed up that I didn't understand what I was doing anymore. I was tired, and I had a bad feeling about what lay ahead. But I couldn't stay in the woods. People were expecting me.

A yellow spot appeared on the trunk opposite me. I turned and, through the branches, saw the lurid moon rising. I sat for another hour and then set out, my stick in hand. The moon was nearly full and shone down like a streetlamp, the trees leaving black pools at the edge of the road.

It was ten miles to Igneada. Well after midnight, I stopped for a break on a ridge facing south, where I could see the lights of the town strung between the darkness of the sea and countryside. It took me another hour to reach the outskirts, where a few hounds barked desultorily. I threw away the stick. A long bridge crossed an estuary, and on the other side there was a campground, mostly vacant. I sat down under a tree, shivering. All I had was a T-shirt and windbreaker. The last hour before dawn was always the coldest.

At first light, I crossed the road to the beach. A rim of crimson widened, grew molten. "Red sky in the morning, sailors take warning," I sang, walking into town.

Igneada was a triangle of kebab shops and hotels, where the inland highway met the coastal road. At the bus station I asked, in pidgin Turkish, for a ticket to Istanbul. The agent wanted my *kimlik*, an ID. I shrugged, and she considered me for a moment, then issued the ticket anyway.

I ate a bowl of lentil soup in one of the shops, and then sat in a café overlooking the bus yard, chain-smoking and sipping scorching thimbles of tea. A truck from the Jandarma rolled up to the ticket office and two cops got out and went inside. The agent must have called them. There was nowhere to run. I tried to prepare myself for what would happen next. The woman would come out and point to me. I would sit, as calmly as possible, as the officers approached and asked for my papers. I was an undocumented Afghan, a refugee, a *mülteci*. They would arrest me. Whatever happened, I wouldn't speak English.

Maryam must have said her dawn prayers by now, so she'd also read the Ayat al Kursi for Omar and me, the Quran's famous Throne Verse. She once explained to me that its power could form walls of protection around a loved one, more transparent than air and stronger than steel. The verse's symmetry led her through the attributes of divinity, of the one true God, who exists eternally in himself, whom sleep cannot overtake, to whom belong all things in heaven and earth, whose will sustains the universe, who knows what is ahead, and what is behind, for nothing can pass except that he wills it, whose throne extends forever, preserved effortlessly, by the highest and most great God.

The cops came out. I watched them drive away and slumped in my chair. A group of passengers gathered in the lot, and when they started boarding, I paid for my tea, walked over, and presented my ticket.

It took us until afternoon to reach Istanbul. I watched the hills give way to factories and shopping centers, and started to feel better. I was out of the woods, and not in jail. But I was in danger of losing the plot. I had to get back to Omar. I promised myself we wouldn't be separated again.

From the bus station in Istanbul, I took a taxi to the address Maryam had given me and climbed to the second floor. Their door was wide open; when I knocked, his sister Farah came into the hall.

"Can I help you?" she said, and then, recognizing me, she laughed.

9

There was still no sign of Omar and Malik in Istanbul. The last Maryam heard, they were in a safe house on the Iranian side of the border, and were about to cross the mountains. Their phones were off, so all we could do was wait.

The family was renting a three-bedroom apartment above a block of sweatshops in Zeytinburnu, a working-class neighborhood in the west of Istanbul. Omar's father had a room to himself while Maryam, her teenage nephew Suleyman, Farah, and one of her friends from Kabul, Shireen, shared the other two, which they'd furnished in the Afghan style, with pillows and mats on the floor. In the evenings, they assembled in the biggest room to watch Bollywood films.

At dinner that first night, as we sat around a tablecloth spread on the carpet, I told them that I'd had some problems getting into Turkey, so I'd entered the country, by swimming across a river. I explained that once Omar got here, he and I would leave together for Europe with smugglers. In the meantime, I asked them to keep my real identity a secret; from now on, I was an Afghan named Habib.

"Habib. That's a good name," Maryam said, smiling, and the others nodded in agreement.

They didn't ask any questions, and I didn't know what else to say. Perhaps they assumed I knew what I was doing, and trusted my intentions. And people who'd lived through war and persecution rarely saw the world in neat categories like legal and illegal. In any case, my arrival was a welcome diversion from the boredom and stress of being stuck here in Istanbul, at the threshold of Europe.

During the day, Jamal stayed clear of Maryam and watched videos on his phone in his room, while Suleyman and Shireen—who, like Farah, was twenty—worked at a nearby restaurant. When Maryam went to pray at the mosque, I was left alone with Farah, who was still looking for a job. She told me about the months she'd spent in Turkey before her mother's arrival, first alone in Istanbul and then with Shireen in the Anatolian city of Eskişehir, waiting for their asylum application to be heard by the UN.

The younger of the two girls, Farah had her mother's broad cheekbones and long, glossy hair which, now that she'd stopped wearing a headscarf, she took an hour to brush before going out. In Iran, her childhood nickname had been *boshke-ye khanda*, or barrel of laughs, and outwardly she was still upbeat, but I'd seen her profile pictures on WhatsApp and Facebook: a bleeding heart shot through with arrows, and the silhouette of a girl, downcast, with WHY ME?? written above.

Like the rest of her siblings, Farah dreamed of emigrating to the West. And like her younger brother, Zia, she'd aced her standardized English test—but there was only enough money to send the boy to study abroad. Instead, she'd found a well-paying job in Kabul with a foreign NGO teaching language courses, and started saving money, thinking she might need it one day for a smuggler, if the Taliban ever came to Kabul.

Then she met Haris. About a year earlier, while she was still living in Afghanistan, he messaged her on Facebook, saying he'd seen her in a friend's wedding video. Like her older sister, Farah was skeptical of traditional marriage, but she liked Haris, who made her laugh. When she finally agreed to meet him, she found him just as handsome as his photos, with his thick hair styled like Ahmad Zahir. A journalist who dressed in sharp suits, he began driving her to work in the mornings, and would often regale her with stories of the powerful people he'd met.

Both families were happy when they decided to get engaged. But the morning after the party, Haris showed up at her house in tears. He had a confession to make: he hadn't been honest with her. She'd known that he was in financial trouble. His father's money-changing shop had burned down, and they owed his clients a lot of money. Now Haris told her that he'd had an affair with an older, divorced woman who promised to pay off his debts if he escaped with her to Turkey. But the woman's ex-husband, a former mujahideen commander, had found out. He and his henchmen kidnapped Haris and beat him until he swore never to see the woman again.

This had all happened before he met Farah, Haris told her, and it was over now. He just wanted to come clean with her. She forgave him. They vowed that no more secrets would come between them. She started helping him out with one or two hundred dollars a month, and soon there was nothing left to save from her salary.

She wasn't in her right mind in those days, Farah would realize later. Her older brothers might scoff, but like her parents, she believed in black magic and the spells that magicians could weave with locks of hair and bits of *haram* flesh. She herself had seen a friend's *khar mohra*, or donkey bead, a hard ovaloid found in the brains of certain beasts from Bangladesh. You could balance the bead on your fingertip and make a snicking noise like you would to call an animal, and it would start to spin on its own. With the right use of a *khar mohra*, you could make someone your slave.

Every morning on the way to work, Haris would hand her a bottle of water and insist she drink. For the rest of the day, it felt like there was a layer of cotton gauze between her and the world. A few times, Haris took her to see a sinister mullah who chanted incantations as she sat with her head bowed, until the room spun. They were spells of protection, Haris told her.

Farah was no longer full of laughter. Depressed and irritable, she fought constantly with her mother and Omar. She'd even stopped seeing her best friend, Shireen. They had played soccer together in the women's league in Kabul; Shireen, short, hotheaded, with broad shoulders and a pixie cut, was one of the best players, and all the other girls wanted to be friends with her, but she chose Farah as her beloved.

They shared everything, good and bad; they had even made the same rows of scars on their forearms. When Farah stopped calling, Shireen was heartbroken but assumed Farah was too happy with her new fiancé to think of her. But one day they bumped into each other, and Shireen noticed how sad her friend seemed. She glanced at Farah's wrist. What had happened to the gold bracelets she always wore?

Shamefaced, Farah swore her to secrecy, and told her that Haris had been desperate; he said his father's creditors were threatening him. She'd already given him her salary, so she sold the bracelets.

With Shireen's help, Farah realized that Haris was taking advantage of her. The last straw was when he'd bought his mother and sisters beautiful new dresses for Eid, after he and Farah agreed not to exchange gifts. Farah broke off their engagement. For weeks, Haris begged piteously for her to take him back, swearing he'd never love another. He waited for her outside her work, and showed her he'd carved her name into his forearm with a knife. One day he forced her to get in the car with him and, as he drove, he pulled his pistol and put it to his temple, raving he'd kill himself.

She wavered, and might have yielded to him if she hadn't discovered—through Facebook—that he was still seeing the commander's ex-wife. When she told Shireen, her friend took her in her arms to comfort her.

"I can't live without you," said Farah.

"I can't either."

"But I can't live here. I want to escape this life."

They decided to escape to Europe together, where they could start all over. It was late 2015, and the border was still open. They just had to get to Turkey and onto the little boats. Maryam, who was getting ready to leave herself, gave Farah some money, as did Omar; Shireen got help from her family. They gave almost ten thousand dollars to a broker who promised them Turkish visas, even though the consulate was mobbed. But after several weeks, the man had come up with only a single visa, for Farah.

The girls had already bought plane tickets; Farah had to fly to Istanbul on her own, before her visa expired. She checked into a cheap hotel to wait for Shireen. But her Turkish visa never came. Finally, Shireen got an Iranian visa so she could fly to Tehran. She lied to her family and said that she was going straight to Istanbul, because she knew they would never let her cross the mountains into Turkey on her own.

Borders concentrate violence, and women who cross them illegally risk assault at the hands of smugglers, guards, and fellow migrants—at the hands of men. Rape is the price some refugees pay to escape. The harsher the border, the greater the risk. The profile of the typical illegal migrant—an able-bodied young man—is often cited to justify strict border controls but it is in fact a consequence of such policies. For the poor, mobility is a part of male privilege.

And yet: Shireen, who kept her hair cropped short, would some-times pass as a boy in Afghanistan, and when she landed in Tehran, she traded her dress and headscarf for jeans and a shirt. She made it halfway across the mountains before her group figured out that she was a woman. Then she felt the smugglers watching her. On the last day, they tried to separate her from the rest of the migrants, saying she was weak, and that they'd put her on a horse—but she yelled and pleaded, and finally the other Afghans banded together and refused to leave without her. And so she made it safely to Istanbul.

The two friends were reunited, but the miracle was over. It was March and the borders in Europe had closed. They didn't have enough money left to pay smugglers to get them to Germany. Once Farah's visa expired they'd both be here illegally. But she had an idea. Farah had heard that the UN was taking applications for resettlement in the West. If they decided you qualified, you could emigrate to a country like the US or Canada. It wasn't easy to be picked but she figured two single Afghan women might have a chance. And while their applica-tion was being considered, she and Shireen would get temporary legal status in Turkey. They got on an overnight bus to the capital, Ankara.

EVERY YEAR, MILLIONS OF PEOPLE flee their homes to escape war. Most don't leave their own countries; they go to another town or a camp, still hoping to return soon. Those who are forced abroad usually don't get far. More than four-fifths of the world's refugees are hosted in the developing world, where borders and humanitarian aid from wealthy countries keep them in place. From this dammed-up pool of the displaced, the West takes measured sips. Out of the more than twenty million refugees worldwide in 2015, around a hundred thousand were plucked from limbo and flown to new lives, the vast majority to the US, Canada, and Australia. Resettlement is a voluntary, humanitarian gesture—in contrast to a country's legal obligations under the 1951 Geneva Convention to asylum seekers who cross its borders—but it also functions as an instrument of soft power, helping to resolve regional crises like *an open shore for an open door* did for Vietnam's boat people. Moreover, signatories to the Convention who make it nearly impossible for asylum seekers to reach their territory, like Australia, can point to their resettlement programs as evidence of their commitment to helping refugees. The right to asylum becomes the privilege of resettlement. Yet while applying is likened to joining a queue, and thus asylum seekers to *queue jumpers*, in truth it's more like a lottery.

In 1979, in the aftermath of the Vietnam War, the West resettled one in twenty refugees worldwide. In 2015, the figure was about one in two hundred. But an individual's actual odds depend upon their nationality, and which third country they apply from. When Farah and Shireen set out for Ankara in the spring of 2016, Turkey was host to the world's largest population of refugees, three and a half million people: Syrians who'd fled the civil war, along with Afghans, Iraqis, Iranians, and many others. But none of these nationalities could claim permanent asylum there because Turkey, for historical reasons, only recognized refugees who came from Europe, of which there were virtually none. Nevertheless, Turkey was at the time a relatively welcoming place for migrants, with its large informal economy and lenient police. Syrians had been given temporary protection by the government; for other nationalities, Turkey had long delegated the process to the United Nations High Commissioner for Refugees, with the understanding that those granted status would eventually be resettled outside the country—and for years, they had. But now the system was breaking down under the weight of new arrivals from the wars convulsing the region.

Between 1995 and 2010, the UNHCR office in Turkey had recognized more than thirty-five thousand Iraqis and Iranians as refugees and resettled almost all of them in the West. Afghan applications, on the other hand, had piled up—the destination countries weren't interested in resettling them. In 2013, in response to an influx of Afghans, the UN stopped registering them. After people sewed their mouths shut and held a hunger strike at the asylum office, the process reopened, but the backlog kept mounting, as Afghans arrived in Ankara to tell their stories.

HER SHOULDER RESTING AGAINST SHIREEN, Farah rode the bus through the night, thinking about the interview that lay ahead, wondering what words would open a door to a new life for her and her friend.

In a refugee camp, stories are everything, wrote Dina Nayeri, who fled Iran as a child. *Every day of her new life, the refugee is asked to differentiate herself from the opportunist, the economic migrant.*

It was early morning when Farah and Shireen got off in Ankara, but by the time they found the asylum office, there was already a long line outside. Most people there were Afghans and Iranians, and Farah kept hearing, amid the Persian conversations, the English word "case." At the end of the line, there was a young Afghan man, clean-shaven, with a child on his hip. They greeted each other and then, looking her over, he asked: "What's your case?"

A case was a story, Farah knew, that answered a question: Why did you leave? She would have to tell it to a stranger who would decide whether or not she was a real refugee. But what to say? A single story cannot contain a whole life.

It is not the voice that commands the story, wrote Italo Calvino, *it is the ear.* According to the 1951 definition, a refugee was someone fleeing persecution on the basis of *race, religion, nationality, membership of a particular social group or political opinion.* When the UNHCR conducted its *refugee status determination*, the interviewer would listen for a *coherent* and *credible* story that demonstrated a *well-founded fear of persecution* on an *individual basis*. War and other collective misfortunes were not sufficient. Afghans were known, among NGO workers in Turkey, to have trouble telling the right story due to cultural norms of modesty as well as their lack of education. For example, an illiterate Afghan farmer who'd escaped from the Taliban-controlled countryside, where his sons were subject to forced recruitment, and his village to airstrikes, might say that the reason he and his family had fled, the last and proximate straw, was because the harvest failed, and there was nothing to eat—in other words, for economic reasons. *Every pain passes, except the pain of hunger*, goes an Afghan proverb.

The right answer to the question of why you left was: Because I was forced. Because I had no choice. But what does it mean to be free in our world? The refugee is freedom's negative image; she illustrates the story of progress that we tell ourselves.

Although the 1951 definition was created for the communist dissident, its application has evolved since the end of the Cold War. In a sense, its scope has expanded, as courts in the West have come to apply it to persecution based on gender and sexuality. But the Convention's protection has been granted to a shrinking percentage of applicants as the number of asylum seekers has grown, in accordance with the more basic law of supply and demand. For example, after an eightfold increase in applications in Sweden from 2015 to 2017, the acceptance rate for Afghans dropped from 74 percent to 38 percent. They became half as deserving.

As single Afghan women, Farah and Shireen would have been virtually guaranteed asylum if they'd somehow made it to Canadian soil. They would have had decent chances inside many EU countries and the United States, but here in Ankara they were competing for a tiny number of resettlement slots. Unless they had an extraordinarily compelling story, their odds were nil.

Farah had a vague sense of this and, in her mind, her ex-fiancé had gone from an abusive stalker to a fearsome Taliban commander whom her family had tried to force her to marry, and who now wanted to murder her to avenge his honor. When she told her case to the young Afghan waiting at the end of the line, he shook his head. "That's not good enough."

"Why not?" said Farah.

"Do you see those Iranians?" The man pointed toward the front, where a group had gathered around a boy who'd come out of the office smiling, as if he'd heard good news. "They're all going to Canada or the US. And the only cases that go there are either Christians, or gays." He glanced at Shireen. "I'll give you some friendly advice. If you want to go to Canada, you should give a lesbian case."

Farah and Shireen looked at each other and laughed. But as the line inched forward, they mulled the man's suggestion.

"What if this is the only way we can go?" Farah said.

"What if this way, they'll send us together?" said Shireen. "Let's try. Please?"

At the last moment, as they reached the front, Farah agreed. They were called in separately for a short preliminary interview. Farah went in first, feeling lightheaded as she followed the Iranian staffer down the hallway. She decided to combine her stories, and the woman nodded perfunctorily through her tale of forced engagement—she must have heard that one before—but when Farah said that she was a lesbian, and was here with her partner, the official perked up. She called a colleague over to confer, and Farah overhead the term "LGBT." Maybe it would work, Farah thought.

Shireen had a similar experience. Luckily, the staff hadn't asked for many details. And to look at them, they weren't the most implausible couple. Farah dressed like a typical Afghan girl, but Shireen continued to wear men's clothing in Turkey. She liked being mistaken for a guy, with her short hair and fake leather bomber jacket, and the swagger in her step. And she definitely wasn't interested in boys like Farah was. The truth was that Shireen's performance of her gender didn't fit with traditional ideas, and back home, that would have meant trouble.

———

Now that Farah and Shireen had applied to the UN, the government would allow them to live in Turkey until their case was decided, but only in the specific city they were assigned—asylum seekers were normally excluded from metropolises like Istanbul and Izmir. The UN sent them to Eskişehir, a leafy university city in central Anatolia. Afghans weren't usually allowed there but Farah and Shireen were assessed to be *vulnerable*; although not mentioned by the 1951 Convention, this quality of vulnerability, based on factors like health, identity, and social group, is today used by the UN and host countries to prioritize refugees for services and resettlement. The anthropologist Didier Fassin argued that such humanitarian logic is the result of *a new moral economy that values suffering over labor and compassion more than rights*. Postindustrial societies no longer desire immigrant labor, and now decide among petitioners on charitable grounds. When Theresa May, then the UK's home secretary, advocated against allowing in asylum seekers, she condemned them as the *wealthiest, luckiest, and strongest*.

Farah and Shireen arrived in Eskişehir without any idea where they would stay, but an Iranian boy they met in the park offered to help them, and within a few days they had found an apartment and jobs washing dishes on Bar Street, a pedestrianized alley that filled up with students on the weekends.

The other refugees in Eskişehir they met were all Iranian; some, like the waitresses at work, were rude to them because they were Afghan, but others were nice enough, like the boy from the park. Though Iran wasn't afflicted by civil war, the Iranians had a better chance of being resettled than Farah and Shireen, because their government was both repressive and an enemy of the West. Some of the refugees were followers of Jesus; though Iran had a traditional Christian community, conversion was illegal for those born Muslim. Western missionary groups were active among the refugees in Turkey; if you got baptized, you might not have to cross the water to Greece. But lesbian, gay, and transgender Iranians were said to be resettled the fastest, like the mullah who had conducted secret same-sex marriages and was featured that summer on the BBC, after he'd fled his country. Farah was told about a lawyer in Ankara who, for a fee of several thousand dollars, could craft anyone a *golden case* at the UN.

The spring passed as Farah and Shireen waited for the phone call that would change everything. In May, Maryam flew to Turkey with Suleyman and joined the girls. But Maryam didn't like Eskişehir. Their apartment was in an isolated housing block, with no shops nearby, and the bus stop was twenty minutes on foot, torture for her knees now that her diabetes started acting up. There was no one Maryam could speak to in her own language; she felt as if she were deaf and disabled. At the mosque, the local women pointedly sat apart from her. Pretty soon Maryam was making Farah unhappy too:

"Why did you bring me here?"

"I didn't! You came here!"

Then Farah heard that her father was on his way from Kabul. The apartment in Eskişehir was too small; her parents couldn't stand to be in the same room together. Maryam wanted to go to Istanbul, where there were other Afghans, and where they could find smugglers. By now they'd learned that the police didn't bother undocumented migrants, if you kept your place. Farah and Shireen talked it over. If they left Eskişehir, the UN might close their resettlement application. But they hadn't heard anything for months, even as their Iranian acquaintances had gotten interviews at the embassies. Omar was supposed to come from Afghanistan soon, and surely he'd figure out a way for them to all go to Europe. And so in June, they all packed their few possessions and got on the bus to Istanbul where, three months later, I would arrive on their doorstep.

10

Not long after I was detained at the Istanbul airport, Omar and his friend Malik had taken a domestic flight from Kabul to Herat, and then a taxi to the Iranian border. Presenting their tourist visas, they continued to the city of Mashhad and spent a night in a hotel near the shrine of Imam Reza, which draws some twenty million pilgrims each year.

The next morning, Omar and Malik took the ten-hour bus ride west to Tehran. It was Eid al Adha, the Islamic celebration of Abraham's leap of faith, when the angel stayed his knife. The holiday would last for several days of feasting and family visits, and their smuggler told them they had to wait until it was over to continue. To pass time, Omar strolled through the park with Malik, marveling at how modern the metropolis seemed in comparison to his visits during his childhood exile. He noticed young couples sitting close together, seemingly unafraid of the religious police. How sweet it was to walk in the evening with your beloved. If Laila were here, he'd buy her an ice cream cone. Impressed by how cheap things were here, he purchased jumbo ones for him and Malik and, for dinner, he had two foot-long chicken sandwiches.

No one bothered them. They passed as Iranian with their fair skin and round eyes, and Omar could still speak the Shirazi dialect of his youth. Besides, they had visas in their passports. There were around three million Afghans in Iran, and their situations varied as much as that of Mexicans in America. Not all of them were illegal like the scrawny kids Omar saw on the street corners, selling trinkets and gum, brands he remembered well.

At sunset, in the warm breath of the road I'll go,
I came on foot, on foot I'll go.
The spell of exile will be broken tonight.
And the tablecloth that was empty, folded.

Mohammad Kazem Kazemi, an Afghan refugee and prominent writer, published his poem "Return" in Iran in the nineties, as a bitter farewell ahead of an imagined departure. Widely read, the poem provoked a moment of soul-searching among Iranian intellectuals over the fate of the Afghans in their midst, the refugees who'd been welcomed in the early days of the Islamic Revolution of 1979, the same year the Soviets invaded Afghanistan. But by the time Ayatollah Khomeini died a decade later, his *Islam without borders* had become an Islamic regime in one country, where Afghan refugees were increasingly unwelcome. Their right to move and work freely was revoked; they were restricted to certain professions, mostly menial ones, and had to apply for permission to travel within the country. *Afghan garbage* were blamed for crime and disease, targeted by immigration sweeps, and deported en masse to their war-torn country. Yet their labor was still desired, and each year hundreds of thousands of migrants made illegal journeys from Afghanistan, forming a precarious, profitable workforce for Iran's construction sites and scrap yards, the farms and slaughterhouses.

And in your courtyard those nights of Eid, neighbor!
You won't hear the sounds of weeping, neighbor!

The stranger without a penny will be gone
The child without a doll will be gone.

There were some Iranians who spoke in solidarity with the migrants, voices which grew louder in moments of tragedy, like the time a local man was able to rape and kill at least forty-three refugee children before the police caught him, since they had disappeared without much outcry. But Iran's immigration laws got stricter; in 2004, the government banned Afghan refugees from most universities and schools. A few days before Omar's arrival in Tehran, the authorities in Shiraz had exhibited illegal migrants in cages in a park. Faced with being deported to a country they'd never known, many Iranian-born Afghans chose to go West instead. When the border in Europe had opened the year before, a steady stream became a torrent as whole families uprooted themselves and headed for the border with Turkey, to cross the mountains on foot.

AFTER EID, THE SMUGGLER PICKED Omar and Malik up in a van and they drove some five hundred miles northwest, to a safe house on the outskirts of Maku, near the border. If they were lucky, the crossing would only take a day or two, but they had to walk through the Zagros Mountains. The previous winter, the sudden increase in migrants had caused the smugglers to use higher passes to avoid overcrowding, and those who'd crossed told stories of corpses frozen beside the trail, of families forced to abandon the elderly in waist-deep snow, of a young mother who, fleeing the police, stumbled and dropped her infant down a crevasse.

At four in the morning, the smugglers roused them and they drove up the mountain in darkness, stopping to pick up two Pakistani clients. When they were dropped off, they followed the trail until they found a group of waiting migrants, guarded by men with pistols in their belts.

The smugglers were Kurds, from a people split across four countries.

The Zagros formed the old, porous frontier between the Ottoman and Persian empires, but the lines between Turkey, Iraq, and Syria were sliced up by the French and British after the First World War. In the bloody century that followed, the borders were mined and militarized, yet the traffic continued across, gasoline and cigarettes, banned cassettes and newspapers, and fugitives from vendettas, coups, and genocide, brought by the qaçaxçi, the smugglers who were the providers in many mountain villages. But it was Afghan opium that made the smugglers kingpins. In the eighties, their heroin labs had started supplying Europe through Bulgaria, whose mafia sold the Kurds weapons in return, and today one of the main trafficking routes still ran across these mountains.

Omar and Malik huddled for warmth as they waited for more migrants, until their group was around eighty people. Most were Pakistanis, young men in sneakers with backpacks, who'd come to Iran through the southern desert and had been traveling for weeks. When it was light enough to see, they followed the smugglers uphill in single file, through reddish hills dotted with shrubs and jagged boulders, along a path lined with flattened cigarette packs. Occasionally, the migrants stopped to take selfies. Below them, the mist in the valley glowed with the morning light.

After two hours of hard climbing, they came over a ridge and looked down into a narrow valley. A road ran along it, with concertina wire strung in the dirt. This was the Turkish border. The smugglers told them to sit and wait while they sent down a scout. After half an hour, the man came back shouting:

"Run! Police!"

As they scrambled back up the slope, Turkish soldiers ran around the bend. Malik, fleet-footed, was near the front of the group and assumed Omar was behind him. When he heard the crack of gunfire, he cringed and looked back, and saw, down below, a slower group of migrants with their hands in the air, covered by the soldiers' guns. Omar was among them; he looked up, spotted Malik, and waved his hands: "Go!"

Three Turks, their bulky rifles in hand, were climbing toward Malik, so he fled upward. At the crest, their smugglers were standing with rocks in their hands.

"*Allahu akbar!*" They hurled the stones at the soldiers, who stopped to shield themselves, giving the remaining clients time to escape.

Omar watched as Malik disappeared over the ridge. The three soldiers came down; one had been hit by a rock and was cursing. The Turks marched their captives back to the road and over the wire, into their own territory, where the rest of the patrol was waiting beside a Cobra armored vehicle. They made the migrants stand in a row.

"*Türkçe? Türkçe?*" one of the soldiers shouted. When the captives stayed silent, he tried again: "English?"

"I speak English," Omar said.

"Afghan?" the man asked.

"Yes," said Omar. The soldier pointed at the rest of the group. "Pakistani," Omar said.

The Turks forced the migrants to hold their arms out in front of them. One soldier picked up a braided metal cable, another a length of plastic pipe, and they walked down the line, whipping their captives on the arm and chest. The more you screamed, the more they beat you.

"Pakistan? Pakistan?" the soldiers shouted. "No Turkey! No Turkey!"

"No Turkey! No Turkey!" the migrants repeated, pleading for mercy in Urdu.

When Omar's turn came, he tried to stay silent as the stinging blows landed on his arms. They passed on to the next victim. Afterward, the soldiers frisked them and took their money and cellphones. Omar was the only one carrying a passport, and the patrol leader came over to look at it.

"Turkish visa?" the man asked.

Omar shook his head.

"Iran visa?"

He nodded.

"Why Turkey?"

"Mother Istanbul, father Istanbul," Omar said.

The soldier noticed the amulet around his neck, the one he had bought in Nimroz, and took it between his fingers. The Throne Verse was written on it. "Quran?" he asked.

Omar nodded. The man gave him back his passport.

The soldiers led them farther down the road, where a minibus pulled up. A man in civilian clothes got down and spoke briefly with the commander, then waved his arm toward the Iranian side. The soldiers laid a wooden board down over the coil of wire, to walk across, and then drove the migrants away with kicks and curses.

Omar and the others ran until they were out of sight of the soldiers, and then started climbing back into Iran. Eventually they saw an Iranian border post in the distance. Omar, who could speak some Urdu, said they should climb higher into the mountains, but others wanted to turn themselves in. In the end, only two of the Pakistanis came with him.

By now it was midday and baking hot as they struggled through the jagged rocks. Omar wondered if he'd made a mistake. They could be kidnapped by rival smugglers, or die of thirst. Suddenly, he heard a voice calling from above.

"Whose travelers are you?" It was a smuggler.

"Musa's travelers!" Omar shouted.

The man told them to keep climbing, and he led them back to the group. Malik ran and embraced him.

"I can't anymore, Malik," Omar said. He sat down, panting with exhaustion. "I'm going back."

"There's no way back. We have to keep going."

IT TOOK THEM THREE ATTEMPTS to make it across. The migrants and smugglers descended to the valley and the Turkish soldiers chased them again, but this time everyone managed to escape back to Iran. In the late afternoon, their lookout reported that the patrol had returned to base. The group went down and crossed the barbed wire. Now the smugglers yelled to start running. They had to make it up the other side of the valley. Omar stumbled up the broken slope of black rocks, his smoker's lungs about to explode. When he tripped, Malik took his bag for him, and together they slogged uphill until they reached the top and collapsed.

By now their group was in bad shape. Many of the migrants had lost their packs fleeing the soldiers, and no one had any water left. Driven by the smugglers, they shuffled onward, dazed, down switchbacks to an arid plateau. In the distance they could see houses. Here their Iranian Kurdish smugglers handed them over to a group of locals.

At sunset, they arrived in a village. When darkness fell they were taken by car into the city of Doğubayazıt. Now that they were inside Turkey, their smugglers were less worried about the police, who mostly turned a blind eye to the passage of migrants. Smuggling was the lesser evil when national security was at stake. The previous summer, a détente between Kurdish rebels and the government had collapsed, and the military had relaunched its counterinsurgency campaign. In the mountains around Doğubayazıt, where the snowy peak of Mount Ararat was visible on a clear day, Turkish units were operating with pro-government militias, some of them linked to smuggling gangs.

The safe house was a single-story building furnished with soiled carpets and blankets, and a fetid squat toilet. The building's door was locked and guarded by a Kurd with a pistol. Until the migrants paid for their journey, they were prisoners. When Omar reached Maryam by phone, she burst into tears of relief. He told his mother to call his friend in Kabul, who would release their payments from escrow. Now they just had to wait until the Kurds in the safe house got their share.

And if they failed to pay? There were three boys there who were being held as indentured workers, cooking meals and cleaning, until their relatives came up with the money. You could try to escape by breaking a window, but you'd be sorry if a rival smuggler caught you. The gangs didn't kidnap their own clients, since that was bad for business, but stray migrants were fair game. Ransoms for an Afghan could reach thousands of dollars, and if they killed you in the mountains, no one would find your body.

Twice a day, clients who had paid were taken in a van to the bus station and sent west. Each night, others took their place, some of them Afghan families who'd crossed with small children, but mostly young Pakistani men. They had less chance of getting asylum in Europe than the Afghans, but even a few years of decent work could be enough to pay for their journey.

"Europe is rich," one of the Pakistanis joked. "Do you know that even in Athens, you can make two hundred euros a day?"

"Really?" Omar said. He'd heard Greece was in an economic crisis. "How?"

"By fucking four Greek men—they'll pay you fifty bucks each."

The others guffawed.

Like war, life on the smuggler's road was mostly waiting punctuated by moments of terror. Long stretches in the safe house were one of the chief miseries. There you were at the mercy of your smugglers for your necessities. In Doğubayazıt, all they were given to eat was broth with potatoes and bread, twice a day. People snored at night in the crowded room. Omar's stuff kept disappearing. While he was in the bathroom, his pants went missing. Someone stole his shampoo. Once he looked out the window and saw an Afghan kid—who'd been thumbing his prayer beads and talking religion the whole time—getting into the van wearing Omar's original North Face boots. Omar pounded on the locked door and shouted, to no avail.

After three nights, the smugglers said their money had come through, and put him and Malik in the next van. To his surprise, Omar saw the safe house was only a stone's throw from the bus terminal. There was already a big group of migrants waiting there; the police didn't bother them.

As a final indignity, the smugglers crammed the bus to more than double its capacity, so the migrants had to take turns standing in the aisle. But they passed without trouble through the checkpoints on the long drive to Istanbul, the Anatolian countryside rolling by, Omar thinking about the boy he'd once been, selling trinkets on the street in Iran; the man without a country he'd become; and the love he had left behind, her face which sustained him onward.

11

Maryam and I watched as Omar and Malik, looking lean and tanned, pounced on the dish of eggs, peppers, and tomatoes that she had made. Omar was narrating his ordeal to his brother in Sweden over speakerphone. "Forget about it, Khalid *jan*, there's nothing left of our feet," he said, swiping his bread around the pan. "They beat us, stole our money . . . No, the Turks . . . Yeah, we took such a beating, I'll tell you all about it some other time. How are you? How's the kid?"

Maryam smiled at me, and I grinned back. I was starting to feel hopeful about our trip, now that we were reunited. But we had to figure out how to go forward.

Later in the day, I went for a walk with Omar and Malik, down our neighborhood's main drag, 58th Boulevard, where a row of storefronts with candy-striped awnings ran south toward the shore. Omar told me that he'd met a smuggler who seemed promising, on the bus from the border; we could call in a couple of days, once he'd rested. We strolled by fishmongers, money changers, kebab and durum joints, and travel agencies with signs in Persian and Arabic. The boys marveled at how modern Istanbul seemed, and how different from Kabul everything was, from the shiny cars that weren't Corollas to the women shamelessly smoking in the street.

"This is nothing," I said. I wanted to take them downtown. "Wait until you see Istiklal Avenue."

As long as you stayed out of trouble, the police didn't care if you were undocumented, especially in a neighborhood like Zeytinburnu, where the people we passed spoke Dari and Uzbek as often as Turkish. Zeytinburnu was the center of Afghan life in Istanbul, which, with its fifteen million, was a city of migrants: Africans and Syrians around the old town, Kurds in neighborhoods vacated by Greeks. Tourists came from all over the world to see Istanbul's palaces and grand mosques, built by an empire that once stretched from Budapest to Baghdad. For centuries, a cosmopolitan ruling class had mixed around the Ottoman court, where a slave from the hinterlands could become grand vizier. When the Allies dismembered the empire after World War I, Mustafa Kemal Atatürk—the Father of Turks—had forged a new nation, with one people under one language, that looked inward to economic self-sufficiency. But after Erdoğan and his Islamist party's victory in 2002, Turkey had embarked on neoliberal reforms that transformed it into a hub of global capital, its economy tripling within a dozen years. Istanbul became home to dozens of billionaires. By 2011, more than thirty million foreigners were visiting yearly. *In the Ottoman period we were a multicultural country, with people of different religion, ethnicity, and culture,* said Abdullah Gül, a former president. *Now again Turkey will be a place with this diversity.*

The optimism of that time was expressed by the hope that Turkey would join the European Union. Since beginning its formal candidacy, the country had abolished the death penalty, decriminalized adultery, and allowed minorities to publish in their own languages. But the eastward expansion of Fortress Europe sharpened divisions with Turkey; when Croatia and Romania entered the EU, Turks lost their right to travel to those countries without visas. And Turkey had become an important route for clandestine migration into Europe; the EU blamed the country's liberal visa policy, which was open to many Middle Eastern and Asian countries not on the so-called Schengen white list. The question of mobility—of who has the right to travel— has become paramount in the twenty-first century. Turkey was Europe's reluctant gatekeeper; the EU wanted the door shut. *When our citizens are insulted on a daily basis in the consulates of EU states,* a

Turkish diplomat stated, *one may ask the question as to why we should help the EU with their problems.*

The past summer, when the boats started leaving by the thousands for the Greek islands, the Turks did little to stop them. In February, the German chancellor, Angela Merkel, came to Ankara to make a deal. In the EU-Turkey Statement the following month, the quid pro quo was made explicit: Turkey committed to stopping refugees, especially Syrians, from reaching Europe; in return, the EU promised three billion euros and a *visa liberalization road map* for Turkish citizens. An open door for a closed shore.

At the same time, the EU announced that migrants would no longer be allowed to leave the islands, and would be held there in camps. Since the deal, the Turks were actively patrolling their coast and arresting migrants. On the Greek islands, a deployment from Frontex, the European Union's coast guard and border agency, was reinforced with border patrol cutters, fast rescue craft, and helicopters. Standing NATO Maritime Group 2, a flotilla that included four navy frigates, was on station *to cut the lines of illegal trafficking.* For now, the forces on both sides of the Aegean were united in their hunt for the little vessels that crossed by night.

ZEYTINBURNU'S BOULEVARD ENDED IN A circular plaza. Across the road, a construction site blocked the shoreline. The fence boards were covered with posters showing glass condominiums in the sky, occupied by smiling, light-skinned couples. We skirted them until we found a passage that led to a park by the sea, where a few Afghan families were clambering on the oily boulders that lined the shore. Omar pulled off his shoes and waded in to his calves, splashing some water on his face and nape, and then sipping some from his palm. He spat it out in shock. "It's salty!" he exclaimed, scrambling out, as Malik and I laughed.

In the distant haze we could see massive ships on their way to the Bosphorus, their decks stacked with thousands of containers. *Zeytinburnu* means "olive cape," and once upon a time its hillside groves were visible outside the ancient walls of Constantinople to sailboats as they tacked along the coast. Now we were deep within the city's sprawl.

"How far is it from here to Europe?" Malik asked.

"A few days, in one of those ships," I said.

He sighed. He needed to stay behind in Istanbul and earn money for his passage. He'd already heard about work in a bakery on the Asian side of the city. It was easy to find an under-the-table, *çabuk çabuk* job here—Turkish for "faster, faster." Omar and I would continue on our own. We had to decide if we still wanted to try to take a boat to the islands. We'd heard horror stories about the camp, that people were eating out of the garbage for lack of food and that violence raged between inmates and police. The week before, the biggest and most notorious camp, Moria, on the island of Lesbos, had burned in a riot.

One alternative was to sail directly to Italy on a smuggler's ship. But it was much more expensive, at four or five thousand euros a head, and riskier, too. If the ship sank, you'd be far out to sea and, if you were intercepted, you might end up in Greece anyway.

Then there was the land route through Bulgaria, back the way I'd come. There you might get beat up, or worse, by the police or vigilantes, and the Bulgarian camps were supposed to be just as awful as the ones in Greece.

I asked Omar what he wanted to do. He was in a hurry now that he'd left Laila behind in Kabul, and the fastest way to Europe was still in the little boats. And it seemed like the shortest line on the migrant's map, in terms of less risk. We were both confident that, once we were on a Greek island, we'd manage to get to Athens somehow—there was always a smuggler's road. "It's your call," I told him.

Omar called the Iranian smuggler he'd met on the overcrowded bus from the border. Yassin had been escorting his own clients, an Afghan family, and ensured they got their own seats, when everyone else had to take turns standing. The smuggler told us to come to a fast-food restaurant on 58th Boulevard, where we found him on the patio. Yassin was tanned, with prominent, tobacco-stained teeth, and wore a houndstooth dress shirt with Gucci-style loafers. There were three others with him; they all got up and shook our hands. Scattered between tea glasses and several overflowing ashtrays was a collection of constantly ringing cellphones.

"I was impressed by how you took care of that family," Omar said. "We're looking for someone who can send us to the islands."

"My business is bringing people from Iran to Turkey, and I hope you'll refer your friends in Afghanistan to me," Yassin said. "As for the boat, perhaps this man sitting across from you can help you."

His companion, who had the sleeves of his crisp white shirt rolled up over his brawny forearms, looked up from his phone and smiled. He introduced himself as Hajji, an Uzbek originally from Sar-e Pol, in northern Afghanistan. "I came to Turkey more than ten years ago, when I was seventeen," he said. "I'm practically Turkish now."

"Back then, he told everyone he was going to Canada," said Yassin, laughing.

"I'll end up there, you'll see," Hajji retorted. When Yassin got up to take a phone call, Hajji gave us his sales pitch.

Our timing was good. Lately the Turks weren't patrolling as much. He wasn't sure why, but just the week before, he'd successfully sent a few boats across. And the Greeks weren't as strict about keeping people on the islands anymore, either. One family he'd sent went straight to Athens after they landed, and another was allowed to go after five days in the camps.

Hajji opened Google Maps on his smartphone and zoomed in on a satellite image of the Greek islands just off the Turkish mainland. We'd wait in a safe house in the port of Izmir until the boat was ready. He'd send us only in good weather, when it was safe. It was a short ride to the Greek island of Chios, he said, expanding the thin strip of water with his fingertips. The journey took half an hour in an inflatable boat, and only ten minutes in a speedboat.

"What's the difference?" Omar asked.

"Five hundred euros for the inflatable, and nine hundred for a speedboat," Hajji answered. If we went in the inflatable, there was a small chance that a Turkish patrol boat would catch us. But even if we were stopped, we'd be detained for only a week, at most, and then let go. We could try again. We could try as many times as we liked, and we wouldn't have to pay until we made it. "The risk and expense are all on me," he said.

Not the risk of drowning, I thought.

"The situation on Chios is good now," Hajji continued. He didn't send people to other islands, like Samos or Lesbos, where it wasn't as easy to get to Athens.

"What if we get stuck on the islands?" Omar asked.

"I'll get you out. Before, when it was stricter, I could get fake Bulgarian papers for 1,200 euros." He smiled. "One more thing—I heard that Canada is planning to take sixteen thousand refugees from Greece. Not just Syrians, but Afghans too. You'd better hurry and go."

Hajji's two henchmen at the table, also wearing pumps and dress shirts, were on their phones coordinating customers, some arriving from the Iranian border and others departing for Izmir.

"What was business like last year?" I asked.

The smuggler grinned. "Forget it! Don't remind me. It was unbelievable. I was sending five hundred people a week. I had three boats going every day. The Turkish police wouldn't touch you once you were in the water. You only had to worry on dry land. I had a friend in the police who was keeping me informed. But boats were too expensive, almost impossible to find. And we got into fights with Turkish smugglers over the best beaches. I beat some of them up. I even beat up some cops, in plainclothes. They said, 'Why did you beat us?' I said, 'We didn't know you were cops!'" He laughed. "In the end, they gave us our own areas."

Yassin, the Iranian smuggler, returned, and suggested we take a walk. Omar and I shook hands with Hajji, and said we'd be in touch.

"Call me on Viber or WhatsApp, not the phone," Hajji said.

We strolled with Yassin through the square, where workers were setting up stands for an outdoor market. He had his own lackey with him taking calls, a bearded kid in a navy three-piece suit, who offered us his pack of Parliaments.

"How much did he ask for?" said Yassin, lighting them for us. "Five hundred? That's too much."

He gave Omar another number and told him to call. The smuggler who answered said he'd send us for 450 euros, and wanted to meet us.

"Forget about Hajji," Yassin said. "I sent some travelers with him and they were caught by the police. I had to suck up to him today because some of my customers are still with him, and I need him to look after them. But he's only in it for the money. That's not the kind of guy I am. Even for a million dollars, I'd never betray someone. You know, when I met you, I could see that you were real men. That's why I want to work with you."

He took Omar's elbow as we crossed the street through traffic.

"Do you think," he said, once we reached the sidewalk, "that you could call your friends in Kabul now, and tell them about me?"

THE ÇABUK ÇABUK SHOPS KEPT their doors open to catch the breeze, and, walking back to the apartment, I could see young migrants bent at tables, the gleam of scissors, loops of smoke around the bare bulbs overhead. Sewing their passage to Europe could take years. Was it better to sit in the cafés and scheme with the smugglers and hash dealers?

There were rumors among the migrants in Zeytinburnu that the border would reopen—if it had happened once, why not again? But no one knew when the hour would come.

Six months had passed and Farah had still no word from the UN about her follow-up interview. At night, she looked wistfully at the ads that came up on Facebook: GET YOUR US & CANADA VISA. She searched for stories about Justin Trudeau, the youthful prime minister who said so many nice words about welcoming refugees to Canada. Maybe she could write him a letter, describing the unfair situation for Afghans in Turkey.

"If you were prime minister of Canada, would you read my email?" she asked me.

"Of course I would," I said, "because you wrote it."

She laughed. "That's not what I meant."

For Maryam and Jamal, there was a slim hope that their sons in Europe could sponsor them to emigrate legally. But for Farah, Shireen, and teenage Suleyman, there was only one way there, the same road that Omar and I were taking. At dinner, Maryam quizzed Omar about the smugglers's rates. Later, he told me that his mother had asked if the girls and Suleyman could come with us.

"What did you say?" I replied, surprised.

"I said my sister can't, because there would be no one to take care of my parents. And Shireen won't go without Farah. But maybe Suleyman can come, I don't know."

"Really?"

"Maybe the police will go easier on us, if we have a kid."

"I think it'll be harder to travel with him," I said. "But you and your family should decide. I don't want to interfere."

I felt sorry for Suleyman, who didn't have anyone his own age around. Before I left Kabul, I'd gone to the village to see his father, Ismail. The farmer had shown me the different varieties of grapes in his field, nearly ripe. "I might have gone myself, if I had been younger," Ismail told me as we walked, his face crinkling into a wry smile. Apart from Maryam, few of his relatives had made the trip to Europe. "We're uneducated. It's hard to have the courage to leave when you don't know anything of the world."

For a farmer to send away his eldest son showed the depth of his faith in education. That was why he'd fostered Suleyman with Maryam in Kabul, and why he'd paid the middleman more than a year's income for his son's visa to Turkey. But their separation wasn't as complete as it would have been just a few years earlier. One night in Istanbul, Ismail rang on Viber—there was mobile Internet in the village now—and we saw the farmer's grainy face on Maryam's tablet, grinning as his son answered his queries: "Yes, Father, I'm well. No, dear Father, I haven't started school yet."

How could he, when they all had to work? Suleyman was missing sixth grade. But he was clever—he'd already learned some Turkish waiting tables. Istanbul was more expensive than Kabul. Maryam's sons in Europe sent her money to help pay rent, and I helped out as well, but the financial pressure added to the tensions simmering in the apartment. Sometimes Maryam scolded Shireen for using too much hot water, or leaving a light on, and the argument would escalate until the girl declared she was going to turn herself in and be deported. Her bags packed, Shireen would sit on the stoop while Farah cajoled her friend and pleaded with her mother. For his part, Jamal was angry that the girls didn't wear headscarves, and that they came and went after dark. But he had scant authority, as Maryam liked to remind him by locking the doors to the other rooms whenever she left him alone in the apartment. She knew he wouldn't dare raise a hand against her. Her kids would throw him out.

Jamal went out to the street at night to escape the apartment, and sometimes I'd join. We looped a shiny outlet mall nearby, the Olivium Center, the old man shuffling in tiny steps as shoppers with bulky translucent bags dodged around us. His back had been destroyed by hard labor in Iran. He could get around on his bicycle in Kabul but here he was nearly housebound. Our conversations began with subjects of mutual interest, like the great tank battles of World War II, but eventually Omar's father would come to what was uppermost on his mind.

"That woman has the Quran in one hand, and the devil's book in the other," he said. "It's a madhouse in there." Jamal told me that, before I arrived, during an argument, Shireen had locked herself in the bathroom with a shard of glass and threatened to slash her wrists. "I had to break down the door. Can you imagine what would happen if she did it? The police would come and arrest me, that's what. The Turks don't like these scenes. They're a calm, orderly people. I should have stayed in Kabul. But I didn't want to be alone."

He'd spent the prime of his life in exile. He had nothing now, not even the respect of his children. They blamed him for not getting them to America. But Jamal's relatives didn't want Maryam there. "Pick two of your children," his brother told him once, on the phone from California. "The two older boys. We'll find you a wife here." But Jamal hadn't gone.

There was no point in regret now. Thinking about the past just clouded his mind with pain. "She turned the kids against me," he told me, as we stood on the stoop of the apartment building. "If I scolded

Khalid for smoking, she'd defend him, and say I did it too. Now he smokes a pack a day. So does Omar." The cigarette fluttered as he held it up. "I didn't want them to be like me."

He slowly mounted the steps. When he had gone inside, I sat down, took out my phone, and started typing the words he'd said. I was emailing myself my notes each night, and then deleting them from my phone. I knew I would one day weld them into a single coherent narrative, but the old man's stories were blurred, details from Iran and Pakistan transposed. They would take force to assemble.

When I finished, I lit a smoke on the stoop. It was late, but I could hear the hum of sewing machines.

OMAR AND I MET WITH more smugglers but we liked Hajji best. At least he had promised to send us to Chios. Above all, we didn't want to end up on Lesbos and inside Moria. When Omar called, Hajji said if he put the money up that day, we could go to Izmir the same night.

Omar and Maryam had a final, private conference while Suleyman was at work. They decided the boy would stay in Istanbul. Omar and I went with Jamal to the money changer that Hajji had recommended. The dim, one-room shop was just off the boulevard, next door to an Afghan restaurant; it smelled like linoleum and there was a murky tank with fat goldfish in the corner. Three men stood behind a case of dusty cellphones. This was a *sarof*, also known by the Arabic term, *hawala*, part of an informal money transfer system that operated on a diasporic scale, much like the Chinese *fei'chien*. You could walk into one and send money cheaply to Kabul, Dubai, or New York, anywhere there was a connection bridged by other *sarof*. Unregulated and mostly invisible to the authorities, informal transfer systems like *hawala* are used by militants and merchants, smugglers and charities, and are vital to the world's undocumented, who are largely denied access to formal banking. No actual currency crosses borders, not for individual transactions; the *sarof* typically partner with merchants back home, so that overall balances can be settled by using remittances to finance imports. As a result, migrant wages flow home in the form of goods, like used cars or cooking oil. That was how much of rural Afghanistan had survived during the nineties.

Omar went to the counter and said that Hajji sent us. The clerk pulled out a thick handwritten ledger, counted Omar's stack of euros, and noted down Hajji's name, the amount, then Jamal's name and number. On a small scrap of paper, the clerk wrote a five-digit code, along with his own cell number, and folded it.

"This is your *ramz*," he said. "Don't let anyone see it. When you get to Europe, tell your family to call and tell us the *ramz*, or else give it to the smuggler. If something happens, if you get arrested, call us before you even call your family, so we can block your money. If the smuggler tells us you made it to Europe, we'll try to reach you. Otherwise, you have three days to let us know if there's a problem before we give him the money."

Because the *sarof* are established third parties who can handle large sums of money, they allow migrants and smugglers to do business in the absence of trust. They are key to the entire underground. The shop's cut was six euros on every hundred that slept in escrow, whether or not we made it. The man handed the slip to Jamal, who squinted at it; it seemed to shrink in his big hand. In the old days, the *ramz* might have been a banknote or Polaroid torn in half and sent by mail. Now transactions happened in real time.

Omar called Hajji, and told him the money was ready.

"There's a bus to Izmir tonight at eight," the smuggler said. "I'll pick you up in the square at seven."

On our way back, we passed the restaurant where Shireen and Suleyman were working. They came out in their red uniforms, the hope apparent on the boy's face that we'd take him.

"It's not happening, Suleyman," Omar said gently.

"Why?" he blurted, and then composed himself. Omar patted him on the shoulder; his turn would come.

At the apartment, we put our few possessions into our backpacks. Omar decided to leave behind his Afghan passport, in case it got lost or stolen; I had no documents. The others hovered nearby, smiling as best they could. It was time to part once again. "I'll go and find my place in Europe, and then I'll see what the best way is for you," Omar said.

"May God protect you," Maryam said. We each embraced in turn, and went downstairs.

"Get some water!" Jamal cried as we reached the stoop, and Farah dashed back up, returning with a potful so that he could toss it behind us, a good-luck charm to hasten our reunion.

It was getting dark already. Omar and I shouldered our packs and set off. The old man insisted on limping with us to the first corner, and then to the next. "It's OK, Father," Omar kept saying. At last, Jamal halted and we left him there, watching us disappear into the crowd.

12

A taxi stopped and the passenger waved us over. It was Hajji. We slung our bags in the back seat and got in. As we pulled into traffic, the smuggler twisted around and handed me a bundle of plastic bags filled with greenish powder: *naswar*, imported from Afghanistan.

"Take it to Izmir," he said. "They ran out, and they've been begging for it."

Reluctantly, I put the package in my bag—I knew it was just tobacco, but it would look like drugs to someone else.

Esenler, Istanbul's main bus station, was a mammoth hexagon of parking decks with the terminal on top. At the entrance ramp, two policemen with submachine guns flagged us down. When they saw our bags, they started questioning Hajji, who answered in fluent Turkish.

"They're Afghans," he said calmly.

The cops looked Omar and me over, and then nodded to the driver. We drove up the ramp and onto the top deck, which was ringed by ticket offices and fast-food stands.

"You don't have any papers, so now I have to bribe the clerk," Hajji complained. He told us to wait upstairs in the lounge, where a Turkish soap opera was playing on TV. Twenty minutes later, the clerk came and took us to where a coach bus was idling. Hajji was gone. The other passengers had already boarded and, as Omar and I took our seats at the very back, the bus rolled out.

We crossed to the Asian side of the city on the second suspension bridge over the Bosphorus, the lights of ferries and pleasure boats dotting the dark water below. To our right, we could see the illuminated span of the other bridge, now renamed the July 15th Martyrs. Two and a half months ago, rebel tanks had blocked the bridge, only to be trapped there by a mob of angry citizens—thirty-four people were killed. But now the city's gloss seemed undisturbed. The billboards and megamalls scrolled by as we passed through a sequence of interchanges and highways. It took us two hours to reach just the outskirts of Istanbul. "It must be one of the biggest cities in the world," Omar murmured. "How much bigger do you think it is than Kabul?"

We drove all night down Asia Minor, the continental seam where past migrations overlapped like ley lines. The written history went back three millennia to when Izmir was founded as Smyrna by Greek colonists, who joined the Athenian empire and fought the Persians. *Do not be afraid of what you are about to suffer,* the prophet John counseled the church in Smyrna under the Romans, after he'd seen his *apokalypsis,* the revelation. The Seljuks conquered Smyrna; the Crusaders sacked it; Tamerlane razed it to the ground. *Very few of those escaped,* wrote the Abbé de Vertot, *who had thrown themselves into the sea.*

When the city grew into a thriving hub for exports to Europe, the Ottomans called it Infidel Smyrna, for its preponderance of Christian merchants, along with the Sephardic Jews who'd fled persecution in Spain. After the First World War, encouraged by Great Britain, the Hellenic army invaded Smyrna, claiming it for the *Megali Idea,* a greater Greece that would encompass much of former Byzantium. After three years of bitter fighting and massacres by both sides, the Greeks were driven out by Atatürk and his nationalists, who expelled one million refugees, part of what was known as the Exchange of Populations. This desperate voyage across the Aegean in 1922 was witnessed by Henry Morgenthau, an American envoy in Thessaloniki. *The condition of these people upon their arrival in Greece was pitiable beyond description. They had been herded upon every kind of craft that could float,* he wrote. *Those who survived were landed without shelter upon the open beach, loaded with filth, racked by fever, without blankets or even warm clothing, without food and without money.*

Many refugees arrived on islands like Lesbos and Chios, which lie like freshly calved bergs just off the coast, and are easily reached by small craft from the mainland, whose forested hills shelter a multitude of coves. When the Nazis invaded Greece, resistance fighters escaped east across the water to Turkey. After the 1980 military coup, Turkish dissidents fled westward to the islands. A decade later, the migrant boats started to cross in war-driven waves, bringing the homeless, stateless, and rightless. After Greece entered the Schengen in 2000, only a mile of water separated Turkey at the narrowest strait from Europe's promise of peace.

WE ARRIVED AT THE BUS station in Izmir early the next morning and had to wait a couple of hours until Hajji's partner, a Kurd, picked us up in a white Hyundai hatchback. He spoke only fragments of Persian, but from his pantomime I understood that he wanted the package of *naswar*. I handed it over, and he mimed wrapping some in rolling paper and putting it in his mouth. No paper, you just put a pinch in your lip, I gestured back. Judging by his excitement, he seemed to think that it was some sort of narcotic. I acted out the effects of taking too much—nausea and dizziness—which only heightened his interest.

We took the highway into town, and then climbed a steep ridge that overlooked the harbor, the houses getting smaller and shabbier as we went. We stopped at a narrow alleyway where the smuggler made a call and a wiry man padded out in flip-flops. He took the package of *naswar*, minus the bag that the Kurd had pinched for himself, and told us in Dari to follow him, so we grabbed our packs and went down the alleyway, where he rapped on the door of the safe house. An Afghan boy opened it, and we stepped inside, our eyes adjusting to the gloom of the tiny studio apartment.

We took off our shoes and sat down on a cot as he hurriedly opened one of the bags of *naswar* and stuffed a wad into his mouth. "I ran out a few days ago," he said after a minute, his voice muffled by saliva.

He leaned back and smiled at us in relief. His wife, a chatty moonfaced woman, came in from the kitchen with a tray of tea, sheep's cheese, and flatbread, and we sat together and ate.

Like us, they were waiting for a boat. The man's name was Sardar, and he'd known Hajji for years. Sardar told us he'd owned clothing shops in both Istanbul and Kabul, and had once been wealthy. But a long illness related to his nerves, and the economic crash in Afghanistan, had reduced his fortunes considerably. He had no faith in his own country's future, and little in Turkey's. There would be a second coup here, he was sure of it, and a civil war would follow.

So in recent years Sardar had been sending his relatives to Europe with smugglers, most ending up in Germany or Sweden. When the border opened last year, he even bought his own inflatable boat and filled it with friends and family. The boat cost five thousand euros in Istanbul, and he paid another three grand in transport and bribes to get it to the coast near Izmir, the main staging point for the great migration to Europe. The city's squares were full of people with life jackets, and on the beaches Turkish TV crews were doing live reports with families scrambling through the surf behind them, onto the boats.

Sardar and his wife's own departure had been delayed by business, but now they were finally going. The kid was his wife's brother, who'd come overland from Afghanistan a few weeks earlier. From Greece, the three planned to go by sea to Italy, maybe inside a shipping container, and then overland to Germany. If they were caught and fingerprinted in Italy, they'd be stuck there under the Dublin rules, but it wasn't the worst place to claim asylum, Sardar said. The Italians rarely rejected Afghans, and you'd get residency fairly quickly.

"That's where I want to go," said Omar.

Hajji had also promised to send them to the island of Chios. The important thing, Sardar said, was not to go to Lesbos, where there'd been the riot and fire in Moria a week ago. He'd paid extra for a speedboat, which was less likely to be intercepted, but Sardar had already been unlucky once. A week earlier, they'd gotten halfway across when a big warship arrived with a spotlight that turned the darkness into day. Their pilot turned back into shallow waters to escape, but there a

Turkish patrol boat stopped them at gunpoint. The Syrians were let go ashore, but the Afghans were taken to a detention camp. It wasn't so bad, Sardar said—the rooms were clean and there were hot showers. A Turkish mullah came and gave a sermon in which he tried to persuade them not to try the crossing again. "Why go to an infidel country?" the cleric told them. "Stay here in Turkey."

When they were released after four days, Sardar and his family came straight back to Izmir. "Sometimes you have to try a few times," he told us. "It's a big game."

"Couldn't you jump in the water to escape the Turks?" Omar asked.

"It's salt water. It would kill you in a few minutes," Sardar replied. He showed us the white stains on his black boot-cut jeans. "Look at how it bleached my pants with just some splashes."

We didn't know when we were leaving; Hajji had told us maybe that same night, but I knew it could take days, or even weeks, depending on the weather and the cops. We'd find out at the last minute, so we had to be ready. But Sardar said that the boats going to Chios didn't leave until the early morning. Omar and I took a nap, and when we woke up we decided to get something to eat. Sardar told his kid brother-in-law to go with us, since he spoke some Turkish. We followed the street downhill, until we found a little restaurant where the staff was laying skewers of minced meat on flatbread piled with herbs and onions stained dark with sumac. We ordered three kebabs and colas.

"This is a Kurdish neighborhood," said the kid as we sat down. "The police don't come here."

At that same moment, a middle-aged cop walked in, his radio and pistol passing at eye level. We went quiet and looked down at our plates. But he was just getting some *adana* kebab to go. And they were tasty: just the right amount of fat in the mince to absorb the cumin and chili.

Afterward, we sat in a park by a dropoff overlooking Izmir. The city was pale and sprawling with a shining stand of skyscrapers at the center. The bay stretched westward, long and electric blue. Omar passed around some smokes he'd bought at the corner store for sixty cents a pack—the filters said Calboro.

"Maybe Hajji will send us all together," the kid said.

Omar and I looked at each other. Were we getting a free upgrade to a speedboat? I liked the idea of crossing with Sardar; hopefully Hajji would take care of him.

Omar and the kid marveled aloud at how beautiful and modern Turkey was, how calm the people were, how no one cared if you wore a miniskirt or hijab or went to a bar or a mosque and whether Afghanistan might one day be similar—maybe in another lifetime, scoffed Omar—and if only you could get papers and open a business you'd have a good life, but then again, Sardar had all that and now he was running away to Europe.

The sun had shattered into opaline shards on the bay. We could see stacks of containers at the port, and the gantry cranes beside them, like giant's gallows. Amid the tankers and ferries, a gray warship coasted into the harbor. I wondered if she was part of the NATO antismuggling squadron, which included both Greek and Turkish vessels, as well as a Canadian frigate, the *Charlottetown*.

On such a sunny day, it was hard to imagine that, in 1922, the quays had been packed with refugees fleeing Atatürk's army as the Christian quarters of Smyrna went up in flames. Some of these migrants continued onward after reaching Greece. *He barely escaped with his life in a small boat crossing the Aegean Sea to Athens and thence on to Ellis Island*, wrote Admiral James Stavridis about his grandfather, a schoolteacher from Smyrna, in his memoir, *Destroyer Captain*. Seventy years later, Stavridis was in command on the bridge of a billion-dollar US warship bound for a port call in Izmir, marveling at the waters his ancestor crossed: *The most amazing historical irony I could imagine is unfolding.*

BACK AT THE APARTMENT, WE started playing *fis kut*, which is a bit like hearts, but only got through a few hands when Sardar's phone rang.

"Get your stuff, they're waiting outside," he told us. So Hajji was sending me and Omar separately. But it was only six p.m., too early for the boats to Chios.

"Oh, you must be going to Lesbos," his wife exclaimed, and then went silent when she saw the stricken expressions on our faces.

Confused, Omar and I quickly packed and went down to the street, where an Arab kid was waiting in the white Hyundai. He brought us to a nearby mosque, where we transferred to a taxi driven by a nervous Turk who took us down a steep blind alley and then had to reverse up, the clutch burning, until finally we found the main road and descended in hairpin turns to the city. The sun had just set; Izmir was a grid of amber mirrored in the water. Omar kept dialing Hajji. When the smuggler finally picked up, Omar asked if he was sending us to Lesbos.

"No, you're going to Chios," he assured us.

I didn't know what to think. Maybe Sardar and his wife were wrong about the timing.

It was rush hour downtown. We stopped in front of a hotel and got out. I looked up: the Susuzlu, three stars. Commuters hurried past. Two more smugglers arrived, conferred with our driver, and then told us to get into an empty van parked there. The smell of leather greeted us when we slid open the door. There were speakers in the ceiling, curtains on the windows, and a mini fridge. It was a party van, moonlighting in the off-season. "Not bad," said Omar, running his hand over the burnished armrest.

Another Turk, tall and bearded, took the wheel. He spoke a little English but claimed he didn't know where we were going. We drove to a residential neighborhood and parked. We waited for half an hour and then the door opened and young men started piling in, one after the other, until they were sitting on each other, their heads at an angle against the ceiling. Omar and I were trapped at the back by a thicket of knees and elbows.

The men were speaking Arabic; I tried the little I knew. The guy on my lap said they were from a village near Aleppo, in northern Syria—I'd been there on reporting trips. They didn't know or care which island we were going to, they were just glad to be out of the safe house after waiting two weeks. "Fifteen days!" his friend said.

"Are there Afghans here?" someone yelled in Persian from up front, in an Iranian accent. "Good, stick with us!"

The driver waited until the escort car arrived, whose job was to drive ten minutes ahead of us, to warn of police checkpoints. When we got on the highway, I saw, to my dismay, that we were headed north toward Çanakkale, in the direction of Lesbos.

"Call Hajji again," I told Omar. He dialed the smuggler.

"Don't worry," he told us, "they're going to turn around."

That didn't make any sense, but what could we do? We were literally stuck. Our van got off the highway and pulled into the parking lot of a shuttered gas station. The escort car up ahead was having mechanical trouble, the driver explained, so we had to wait. He told us to shut the windows and curtains and keep quiet, then took off, no doubt afraid to wait with us in case the police came.

The van soon grew stiflingly hot. The Syrians whispered to one another in the dark. My joints were throbbing and my neighbors' skinny limbs pressed into my kidneys. I felt a wave of nausea, and closed my eyes. As a kid I used to have nightmares about being trapped in a dark space filled with pressure and heat, like the center of the earth. Back when I was planning this trip, I read about seventy-one migrants who had suffocated in a meat truck in Austria that past summer, and promised myself to never get into a situation like that. Now I thought about what the man in Nimroz had told us: Brother, we're like a football, being kicked up the field. There was a sense of vertigo in handing yourself over to criminals. No recourse, legal or moral, for what befell you; the blame, rather, for putting yourself there in the first place.

An hour passed like that. It was almost ten o'clock when the driver got in and we set off again, still heading north. As had been obvious for some time, Hajji was lying. Omar called him again and they argued, and then Hajji told him to pass the phone to the driver. Up it went through the tangle of limbs.

"OK," Hajji admitted when the phone was relayed back, "you are going to Lesbos, but what's wrong with that? It's a fine island."

"We don't want to be locked up in Moria!" said Omar.

Hajji said he'd send us back to Izmir with the driver once we dropped off the other passengers, and then hung up.

"What should we do?" whispered Omar. We were crammed tête-à-tête.

"What can we do?" I said.

"Nothing. Just nothing."

After forty minutes, we exited the highway and, joined by two other vehicles, went down a series of increasingly deserted country roads, moving fast through the olive orchards and sleeping villages. We came down a ridge into the narrow streets of a coastal town, Bademli, and drove through a pine forest with the headlights off. At the beach everyone jumped out except for me and Omar. We sat there and tried to explain that we were going back to Izmir.

"Go, go," hissed the driver. "Police, police!"

Another stocky smuggler came and yanked me out by the arm. Behind me, Omar kept resisting until the guy pulled a pistol out of his belt. Then Omar got down quick. The vehicles took off, leaving behind three van loads of migrants, two Arab smugglers, and a deflated rubber dinghy.

Omar called Hajji.

"They left us," Omar said.

"What? OK, I'm sending someone." The smuggler hung up.

"Forget it," I said. "He's not sending anyone."

"That *khar kos lauda!*"

We sat on the ground, stunned. What could we do, besides try to escape into the woods? We might get shot that way. Like it or not, we were going to Lesbos.

"Here," Omar said, handing me an empty garbage bag. I put my pack inside of it and tied it shut. We were at the end of a long, narrow inlet, where the road came up to a ridge of beach grass, and then curved back into the pines. In the darkness, I could make out around forty people squatting in the tall grass, some of them women and children, the youngest infants. The majority were Arab; there were a few Africans as well. Most but not all were wearing life jackets; some also had inflated inner tubes.

"Where are our vests?" Omar said.

Back in the safe house, we'd been given the option of buying a life jacket, but even though we'd forked over thirty-five bucks each, the smugglers now told Omar they'd all been distributed. I wasn't going to argue, but Omar got mad and started yelling that we weren't getting in the boat without them. The smugglers hushed him and one came back after a minute with a single jacket.

"That's all there is," the man said.

I told Omar to take it.

As we were arguing with the smugglers, the two Iranians from the van crowded in, anxious to ask us something. One stood over me as I prepared my pack, repeating himself in his thick Tehrani accent, as if he couldn't understand what I was saying.

"Are you Afghan?"

"Yes."

"Are you Afghan?"

"Yes!"

"Are you Afghan?"

I leapt to my feet and shouted at him: "I said I was Afghan, are you deaf?"

Omar separated us. The other migrants were staring in alarm.

"Yes, we're Afghan, don't worry," Omar said to him.

"Look, we're going to say that we're Afghan too, OK? Now tell me, who is the president of Afghanistan?"

I got up and walked away, trying to calm down. There's two presidents, I thought, and then laughed. The Iranians wanted to pass as Afghan for the authorities in Greece, who sometimes asked questions like that. Hopeless. But I was losing my cool; I shouldn't have yelled.

Omar and I went down the beach, where we'd have a better chance of making a run for it if the police came. The smugglers had forbidden smoking, but we lit a cigarette anyway and shared it, shielding its ember in our cupped palms, as we stared at the black surface of the inlet.

"Are you ready?"

"I am, brother."

Wheeze, wheeze, wheeze. The smugglers were inflating the dinghy with a foot pump. Now and then a car would come crunching down the gravel road, and everyone would hush up and crouch, the headlights silhouetting their forms for a moment.

How many had come here before? I was sure we were neither the first, nor the last. Like us, they had left behind what they could not carry. On this beach, they had prayed in different tongues for the same thing: to pass across the water.

Now a group of refugees came down the slope, bearing the boat's green and white pontoons on their shoulders. The rest of us gathered as they launched the dinghy, and watched the smugglers fit on the outboard engine. They waded the boat out waist-deep, and gave the helmsman some final instruction. He was just a refugee they'd selected, most likely in exchange for a free ride, since the inflatables made one-way journeys. When the smugglers signaled to us, we grabbed our bags and rushed forward.

The cool water on my calves sent a jolt up my spine. When I reached the boat, I hauled myself up over the squeaky rubber and found a seat on the port side. Omar got in next to me.

"One by one!" the smuggler said in Arabic. After the men had lined up on the pontoons, facing inward, we put the women and children on the floor in the center. There were almost fifty of us packed into twenty-five feet of rubberized canvas.

The starter cord snapped and the engine roared to life. The smugglers gave us a shove and the helmsman clicked the outboard into gear, a long V rippling out behind us. On an island at the mouth of the inlet, there were stone buildings and wood cabins, lit up in mauve and green: a luxury resort. It was near midnight, and a guest out late would have seen us just offshore, huddled and riding into darkness.

As we rounded the point, a shimmering band came into view on the horizon: Lesbos, about ten miles away. The sea loomed around us as the land shrank away. The boat began to rock in the swells, which had the choppy motion that comes after a storm. Apart from the floorboards, the dinghy lacked any structural support and undulated with each wave, pitching us against one another. Our vessel was laughably unseaworthy, an overgrown pool toy ready to pop.

We had to reach Greek waters before the Turks caught us, but the helmsman, a young Syrian with a fuzzy lantern jaw, kept the throttle at idle, despite the urgings of the passengers to hurry. Perhaps the smugglers had warned him not to overstrain the dinghy. He stood in an unsteady crouch, swinging the tiller back and forth, and each time we dipped into a swell, he called out an invocation to God, like a whistle buoy: *"Ya Rab! Ya Rab!"* His friends beside him were making a racket too, speechifying in Arabic about having faith and leaving Turkey and more that I couldn't follow.

A curly-haired little Iraqi girl was sitting with her parents on the floor in front of me. As the swells grew rougher, her head kept knocking against my knee, so I reached out and cradled her head. Her mother didn't seem to notice; she looked like she was about to be sick. It was too dark to see the other passengers' faces clearly, but as I listened to their whimpers and groans, I became aware of the utter terror that surrounded me. I could feel Omar tensing beside me with each wave. It was the first time that he'd ever been to sea; the same must have been true for many of my boatmates. We are most frightened of the unknown. It was different for me. I'd grown up on the ocean. I felt another kind of dread. I imagined the water rushing inward, with its sudden chill, and rehearsed what I'd have to do. I was a strong swimmer and I could save myself. I would help Omar, but I knew that without a life jacket I'd have to get away from the others fast, if I didn't want to get taken down in the panicked thrashing. We'd swim clear and wait. The drowning are silent; you need air to scream. Afterward, for those on the surface, it would be a matter of time. I'd heard that the jackets were often fake and would absorb water. Even if we stayed afloat, we'd eventually die of exposure although the sea was likely warm enough this time of year to survive through the night. It was five miles to the shore from midpoint, and there were currents—a long swim. Better to stick together and wait for rescue, if it came.

The girl's skull was warm on my palm, and I realized that she'd fallen asleep. The moon had yet to rise, and stars danced in the facets of the waves. Red and green lights blinked up and down the coast. We were crossing Europe's moat. The Mediterranean is the world's deadliest border. Since the year 2000, more than thirty thousand migrants were recorded as having died there, alongside an unknown number. *Nature sanitizes the killing floor,* as Jason de León wrote about the Sonoran Desert and its vultures, between the United States and Mexico.

The loss of the body means someone cannot be properly grieved—thus the horror of drowning in so many cultures, and the ghosts who wait by shore.

THE WARSHIP APPEARED SILENTLY, AS in a dream, her jagged silhouette black on midnight blue. It was a NATO frigate, maybe half a mile out, steaming parallel to our course. She must have followed our radar signature here, or else a blotch of heat. We watched quietly as the ship drew closer. When her blinding searchlight came on, the passengers cried out, some weeping, and others praying. Across from me, a diminutive Eritrean man made the sign of the cross, his irises catching the glare as we looked at each other.

"Go, *sharmuta!*" yelled an Arab boy at the helmsman who still refused to speed up.

Earlier, I'd spotted a red-and-blue light flashing far astern, a vessel passing south along the Turkish coast. Now the boat changed course to a fixed bearing, approaching and entering the spotlight's ring, a white rigid-hull inflatable not much bigger than our own craft, with SAHIL GÜVENLIK on the side. The frigate must have alerted the Turkish coast guard, which had come to arrest us and take us back.

There were two men at the bow of the Turkish boat, carrying flashlights but no firearms that I could see. The passengers went silent for a second and I figured we would go peacefully. As the Turks closed in, one started yelling in English, "Engine off! Engine off!"

But our boatmates rose to their feet and shouted back: "No Turkey! No Turkey!"

Then they rammed us hard amidships, sending up a shower of seawater and pitching me forward into the woman and her child. The Turks were trying to pin their pontoon against ours so they could take control of our vessel, but our helmsman revved the throttle and we jerked away, people whistling and jeering.

The warship had turned off its light and faded away into darkness. The coast guard looped around and then roared in again, faster this time. I had to jump up to avoid getting crushed as we collided. The Turk put the wheel over to force our bow around, but our helmsman kept goosing the throttle. Beneath the screaming engines I could hear the thump of limbs against the pontoons, the hiss of water from the bow. One of the Turks tried to throw a line around our outboard, but the Syrians tossed it back and threw punches, too. The other Turk went at them with a boathook, slashing it against our pontoon. Our helmsman hit the throttle and we shot away, our boat chanting: *"Allahu akbar!"*

It was at this point, I later learned, that we crossed into Greek waters. The Turkish boat was circling back when, a couple hundred yards to starboard, a larger boat revealed herself by turning on her deck lights. The passengers stared as our helmsman slackened the throttle.

"It's the Greeks!" someone said.

"No, it's the Turks, keep going!" shouted others. The other vessel, closing in fast, fixed us with a spotlight.

"Show them the children!"

The mother at my feet stood and, braced by the others, hoisted her curly-haired child aloft, the girl blinking back tears as she gazed into the light. There was a buzzing sound; I turned and saw the Turkish boat bearing down.

"Look out!"

They hit us so hard that we bounced away; the mother and her child were knocked over into the tangle of bodies. Were they trying to sink us? Again they looped around, but this time, the other ship, a big cutter, swung in between us and the Turks broke off their pursuit. Our dinghy was still running hard for Lesbos; now the black triangle of the cutter's bow loomed over us. Burly sailors with pistols on their hips stood at the railing, backlit, yelling down in English:

"Stop the engine!"

There was no way we could escape, yet my boatmates argued over whether to surrender. Then we saw the illuminated flag flying from its quarterdeck: the blue-on-red cross of Norway.

"It's the Greek flag!" they shouted joyously, and broke out in ululations. The helmsman cut our throttle and the ship pulled alongside. The refugees, agitated, kept shouting "No Turkey! No Turkey!" up at the Norwegians, who tried to calm us down. Omar went astern and spoke to our rescuers in English; they asked everyone to sit and stay quiet. Finally, one by one, women and children first, we were hoisted onboard, frisked, and sent forward on the open deck at the bow of the cutter.

She was the *Peter Henry von Koss*, a rescue vessel detailed to the Frontex mission. As we got under way, the wind began to slice through our clothes, which were soaked from the melee. Under the deck light, I could see my boatmates clearly for the first time. The terror was ebbing from their faces, replaced by an exhausted triumph: they had gambled with their lives against the sea, and won. Whatever happened next, we had made it to Europe.

The Norwegians came forward with thermal blankets for the families; they even had teddy bears in life jackets, for the kids.

"You are nicer than the Muslims," one of the Syrians said in English, his voice thick with anger at the Turks.

A police officer with a fleshy, genial face counted us and then radioed in the numbers by nationality. When he spotted Omar, he squatted down beside him.

"Thank you for your help," he said. "How come you speak such good English?"

"I served with the coalition forces in Afghanistan," Omar answered. "I'm escaping the Taliban."

The Norwegian explained that Omar would have to wait for months in the camp on the island while people were sorted out. Many migrants were coming in search of a better life, the man said, and Europe didn't have room for all of them. They wanted to take the place of real refugees like him.

"You know, I think you have a very good chance of making a life for yourself in Europe," he said.

"Thank you, sir," said Omar.

"You're acting like a European. You're being calm and quiet. Europeans don't like it when people start shouting and acting crazy. Why was everybody so worked up? It's very unusual for us." He gazed down at us, perplexed.

Omar sighed. "It was because of the fight with the Turks."

The Norwegian asked him how much he had paid for the trip. "You know," said the cop, writing the answer in his notepad, "the smugglers indirectly kill people by putting them in these unsafe boats. Like these green ones. We've seen some of them lose their air halfway across. Everyone goes into the water, and people die."

"What will happen to us once we get to Lesbos?" asked Omar.

"You will be taken to Moria," he said, and frowned. "The Greeks aren't going to be as nice as us."

We were shivering by the time we arrived in Mytilene, Lesbos's main port. It was four in the morning. My boatmates lined the rail to see the stone buildings along the harbor. "Look, a mosque," one of the Iranians said, pointing at an illuminated church steeple.

The Greek police were waiting as we disembarked on stiff legs. The Norwegians had already piled our bags on the dock, and when I went to take mine, one of the Greeks grabbed me by the arm and shoved me, barking at us to form ranks. Once we were lined up, a cop explained in English that the registration office was closed until morning, so we'd have to wait in the port. The families were taken to a tent and the rest of us were marched onto an empty bus.

Omar and I sat together near the front. He passed out on the hard bench seat immediately, but I sat there in my wet jeans, too keyed-up to close my eyes. My phone was dry and still had some battery left, so I wrote down what had happened. I still had a sense of motion from the waves. The faces I'd seen in the darkness were alive. But I didn't feel joy, just worry for what awaited us in the morning, when we would be taken to the most notorious camp in Europe. When I'd first thought to follow Omar, the border had been open and I figured we'd land unnoticed amid the masses and continue onward. I didn't want to go into the system. What if we couldn't find a way to escape from Moria?

At eight, another bus arrived, followed by a Dutch prison van. We transferred and drove through Mytilene's streets, past a wall graffitied with NO NATO NO FRONTEX, and then north along the coast. After a couple of miles we turned inland, through rolling olive orchards. We saw a team of riot cops in tan fatigues, and then arrived at a compound with guard towers, fenced with razor wire.

Part III

The Camp

13

On every island, the people watch the sea. On Lesbos, spring brought the little boats. If you stood on the northern headlands, you could see them coming from the Turkish side of the channel, dots that grew larger until you could make out the bright life jackets and hear the whine of the engines, then the people's cries as they came through the surf. They landed on the beach if they were lucky, on the rocks if not.

The boats arrived like messages from distant wars; some years brought more than others. In 2014, as Syria burned, some forty thousand people landed on the Greek islands, the most yet. Over the winter, when the sea turned stormy, only the most desperate made the crossing. But the following spring, of 2015, when the weather warmed again, it became clear that this would be a year unlike any other.

That March, almost eight thousand came ashore on the islands, an unusually high monthly number. The flow doubled in April, and kept increasing so that fifty-five thousand landed in July. By then, the world was waking to the crisis, and the media flocked to Lesbos, the photographers stalking beaches turned orange with discarded jackets. Their cameras watched as the boat people flung themselves to the sand and wept with joy, thanked God and took selfies, and then started walking. At the overwhelmed registrations center, the Greek authorities handed out pieces of paper ordering the migrants to *self-deport*, and then let them board the ferry to Athens, where they headed north through the Balkans, into Europe.

In August, a hundred thousand people landed in the Greek islands, most of them on Lesbos. A majority were Syrian, and there were unusual numbers of women and children among them, as it became apparent that you could travel safely once you reached solid ground. By December, a third of those coming ashore were underage. Not all made the crossing: of the 800 dead or missing at sea that year, 270 were children.

It was pandemonium on the beaches. Some days it seemed like the end of the world: *And the sea gave up the dead which were in it.* Corpses washed up without head or limbs; others were untouched, as if asleep. The fishers' names for the shore changed, one resident would later write. A rock they'd called the Seal was now the Old Man. Some locals refused to catch fish, saying they were *fed from the corpses of drowned immigrants.*

On September 2, a family of four got into a boat in Turkey. They were Syrian Kurds from the city of Kobani, where, three months earlier, ISIS had massacred more than two hundred people. In the predawn darkness, the boat capsized in rough seas; the mother and her two boys drowned. At sunrise, the three-year-old—Alan Kurdi, or Aylan as he would be known to the world—was found on the beach. A Turkish photographer shot the scene: a child collapsed forward the way exhausted toddlers will, his red shirt hiked to show his white belly, a wave lapping his brow. He still had his shoes on.

There were many pictures taken of drowned children that summer, but perhaps the boy's angelic repose was why his image went viral on Facebook and Twitter, at a rate of more than fifty thousand tweets per hour, reaching an estimated twenty million screens. *What happened to Aylan Kurdi?* was a top related search on Google, along with *What is causing the migrant crisis?* His body appeared on the continent's front pages: *An image that shakes the conscience of Europe,* headlined *El País* in Spain. The UK's bestselling paper, the *Sun,* whose columnist had written *these migrants are like cockroaches* five months earlier, announced a charity campaign entitled *For Aylan*. The top tabloid in Germany, the conservative *Bild*, displayed the child's corpse on a stark black page. *This photo is a message to the whole word to finally unite and ensure that not a single child dies again on the run*, read the caption. *After all, who are we, what are our values really worth, if we continue to allow this to happen?*

The artist Ai Weiwei visited the beach on Lesbos and lumped himself in the same position as the boy for a photo. Accepting a prize in Berlin, the writer Karl Ove Knausgaard said the image had shocked him into paying attention to the refugee crisis, which had hitherto blended in with *the constant reports of car bombings in Iraq or school shootings in the United States*. Senator John McCain, who'd urged the US to send troops into Syria, posed with a photograph of the dead toddler in Congress.

From a child, the miracle. On the second day after Aylan's death, Germany and Austria opened their frontiers to thousands of people bottled up in Hungary, letting them take buses and trains northward. Other countries followed suit, forming what was dubbed a *humanitarian corridor* through the Balkans. On Twitter, #refugeeswelcome became a trending topic. In Munich, ecstatic crowds bearing chocolate greeted the arrival of migrants; in Vienna, hundreds of rail workers offered to work overtime for free. On September 10, the German chancellor, Angela Merkel, posed for a selfie with a Syrian asylum seeker. *We can do it*, she'd announced.

The border was open. On Lesbos's beaches, you could find cops, anarchists, tourists, and missionaries working side by side to help migrants through the surf and onward to Europe. In October, two hundred thousand people landed on the Greek islands, a flow that peaked at ten thousand in a single day. One million crossed in fourteen months. It was the greatest movement of refugees by sea in history.

But as winter approached and the waves grew rougher, the mood in Europe was also changing. *It's not like a month ago when lots of ordinary people would come and applaud when they arrived*, said a cop in a German border town that October. Most Germans had approved of Merkel's decision to open the border; now a majority said she'd been wrong. In truth, Europe's leaders were working frantically to stanch the flow of refugees; Austria, Hungary, and Slovenia were building razor-wire barriers along the Balkan corridor.

On November 13, nine men with assault rifles and suicide vests attacked Paris, killing a hundred and thirty people. Two of the terrorists, it would emerge, came through the Greek islands by posing as refugees. On New Year's Eve in Cologne, amid the crowds packed into the central train station, hundreds of women were sexually assaulted and robbed by groups of men. On January 1, the Cologne police had described the *exuberant* evening as *largely peaceful*; three days later, in the face of a mounting outcry on social media, the police chief announced that crimes of *a completely new dimension* had taken place and that the suspects were Arab. The tabloid *Bild* warned of *sex mobs across Germany*. The French satirical weekly *Charlie Hebdo* ran a cartoon of the drowned boy, entitled *Migrants*, which asked: *What would little Aylan have become if he'd grown up?*

The answer, beneath a drawing of men chasing terrified women: *An ass groper in Germany.*

We must now secure Schengen's external borders, Wolfgang Schäuble, the German finance minister, told the press. First, the land borders were closed with barbed wire and riot police; then, on March 18, the EU-Turkey deal to stop the boats was announced. Asylum seekers who made it ashore would be detained on the Greek islands and processed by a *hotspot system*. The linchpin and largest camp was on Lesbos.

OUR BUS DROVE THROUGH MORIA'S main gate and stopped at an inner compound, also fenced with concertina. Out the window, I could see a line of officials standing with clipboards, some in UN vests. As I got down with the others, my legs felt wobbly. I didn't have to do this, I thought: I could just cry out in English that I was a journalist and that I'd been tricked by the smuggler into coming to Lesbos. If I did that, not only would they separate me from Omar, but if word got back to the smugglers there'd been a spy among them I might put him and the other migrants in danger. Anything could happen in the camp. I stayed in line.

As I walked through the gate, a woman fastened a paper bracelet with a serial number around my wrist. We were led into a waiting area with some wooden benches under a canvas hangar, where we were seated according to language: the Arabs were the largest group, followed by some Tigrinya speakers from Eritrea. Omar and I sat with the Iranians, as interpreters in vests translated for a Greek official. The man explained that we'd be fingerprinted and given a medical exam, then sent to the main camp. We weren't allowed out for the next twenty-five days. Once our registration was finalized, we could visit the rest of the island, but we couldn't leave Lesbos until we passed our first asylum interview. And we weren't going to travel onward to other countries in Europe. If we didn't like that, we could go back to Turkey.

First they called the Eritreans to get registered. Afghans were next. The two Iranian boys had changed their minds about trying to pass, so just Omar and I entered the administrative compound, where a planter in the courtyard held three windswept pines with whitewashed trunks. The container offices, trimmed in Aegean blue, were arranged in a counterclockwise circuit; our first step was the *nationality screening*. An officer from Frontex, the EU border agency, would try to verify whether we were really Afghan.

Because some nationalities were given favorable treatment by the authorities, Iranians or Pakistanis might try to pass themselves off as Afghans, just as Moroccans would as Syrians, or Ethiopians as Eritreans. Since, like Omar and me, many people traveled without documents, and fakes were readily available in Turkey, the Frontex screeners and their interpreters would quiz people about their home countries. They used to ask simple questions about flags or currencies, but when the migrants got wise—some smugglers even sold cheat sheets—the screeners tested people on landmarks back home, and how many digits there were on license plates, or what the price of flour was. The trouble was that many refugees had spent their lives outside their own countries. An Afghan who'd grown up in Tehran spoke with an Iranian accent, and probably knew less about Kabul than I did. Decisions were quick and summary. Mistakes could be made.

Omar went in first, and I sat down on a bench outside. Some of the children from our boat were playing on a toy set, while a couple of short-legged mutts sniffed around. It was almost nine o'clock, and the staff was arriving with cups of coffee in hand, some in uniform, Dutch prison guards and Greek coast guard officers. Two men in cargo pants chatted in French while chain-smoking by the fingerprinting office. Then there were the familiar NGOs: Mercy Corps, ActionAid, Save the Children. The aid workers were younger than the officials, and reminded me of the kids I'd seen rotate through Kabul: slim hips, expensive boots, airs of concern. My boatmates wandered among them bewildered, their registration papers in hand. Several containers had the same poster of a white hand pulling a brown hand from the waves. UN ARRÊT ICI, it read: *A stop here.*

A Frontex officer with spiky hair and glasses stepped out of the screening office: "Afghan?"

I followed him inside and we sat down at a round plastic table, my pulse hammering in my ears. A middle-aged, bored-looking Afghan interpreter was already seated, and we greeted each other in Persian. He translated as the Frontex officer ran through a list of basic questions. No documents? No problem. The officer mostly wanted to practice the bits of Dari he'd learned.

"*Az kuja hasti?*"

"*Ma az Kabul umadom.*"

"Where in Kabul?" the interpreter asked sharply.

"*Qala-e Fatullah.*"

"OK," said the officer, scribbling at the bottom of the form.

"That's it?" the interpreter asked him in English, surprised.

"That's it," said the officer, and then smiled at me.

"*Khoda hafez.*"

"Tell him thank you very much," I said, and walked out with my form, blinking in the bright daylight. So it was official. I sat on the bench and had a cigarette, feeling drained of emotion. The day was warming up and as I sunned the damp crotch of my jeans, I wondered what had happened to Omar.

In fact, it was him Frontex took an interest in. A Belgian investigator overheard Omar talking in English about how he'd worked as a translator for NATO and asked him to come for a chat in another container, where they were joined by a Greek cop.

The friendly pair offered Omar coffee and the WiFi password, and then asked about his smuggler. They were especially curious how payment worked. If Omar wanted, he could use their phone to call his contact, to let him know his boat had arrived. Could he give them the smuggler's number?

Omar was chatty but evasive. The smugglers used fake names and they were always changing their numbers, he said, so he hadn't bothered to remember them. The investigators said Omar would be stuck in the camp for a while, and that they could get him work as an interpreter. He replied that he'd think about it.

When Omar got out, he joined me in line outside the fingerprinting office, our next stop on the circuit. A central function of the hotspots was to enroll migrants in the European Dactyloscopy Database, or EURODAC. This was used to enforce the Dublin treaty's requirement that asylum seekers apply in the first EU country they reach, and then stay there. In practice, that meant the first country where you were caught, hence the importance of staying underground with smugglers until you reached your destination—some migrants, upon arrest, even burned or mutilated their fingertips so they couldn't be scanned. Otherwise, if you applied for asylum in, say, Germany and they found your prints in EURODAC showing you'd been arrested in Italy or Bulgaria, the Germans could deport you back there. But not to Greece, not since 2011, when the European Court of Human Rights had ruled that conditions were too atrocious here. The EU was trying to fix the situation, but for now, Omar had nothing to fear. I let him go first.

Through the open doorway, I watched as one of the Frenchmen I'd seen earlier, now wearing latex gloves, rolled Omar's fingertips and then his palm onto the glass of a Crossmatch scanner. Now Omar would be legible to the state. When EURODAC became operational in

2003, the database was strictly limited to asylum uses. But in 2015, the rules were changed to grant law enforcement access for *serious crimes and terrorism*. That was how the police matched the prints of two of the Paris attackers who'd landed on the island of Leros posing as refugees, dead men who were never identified. Now EURODAC was being merged with other databases into one vast system that would track all visitors to the EU. On the official's screen, I could see the whorls spidering into an image of Omar's fingerprints, and then the program beeped to indicate a clean capture.

Innovation happens at the periphery as often as the center; fingerprinting was developed by colonial officials in British India before it was adopted by Scotland Yard. In our day, technologies like biometrics, drones, and automated mass surveillance move from overseas wars to the borderlands and finally the metropole. In Afghanistan, my prints had been recorded, along with my irises, by the US military prior to my embedded reporting trips, part of an immense trove collected there and in Iraq that was shared with the FBI and Homeland Security.

When it was my turn, I walked up to the desk and stood facing the French official, who lined up his camera to take my picture. I hadn't yet had my prints taken on the continent, but I wondered if, post-Paris, the Europeans were secretly running checks with the Americans. And if they didn't discover my real identity, would Habib, my alter ego, come back to haunt me? When flesh became ones and zeros, your double could live in the cloud forever. But there was no way I could avoid this without endangering my friend. The official took me by the wrist; the scanner glowed greenly.

Biopolitical tattooing, the philosopher Giorgio Agamben called it. The man pressed my hand to the glass.

"BROTHER, IF THEY FIND OUT you're Canadian, it will be very funny," Omar said as we filed back into the hangar during the lunch break. He laughed. I squinted at him through the haze of my fatigue, trying to determine whether he'd lost his mind. Maybe he was just happy to be alive. I chuckled, too.

Through the fence, we could see the inmates in the main camp passing by, dressed in motley, donated garb. Spotting us, some Afghans called to us through the chain-link.

"Salaam! How many Afghans are there with you? Any families?"

They were looking for relatives who were supposed to have made the crossing. Loved ones they hadn't heard from. Names and descriptions we didn't recognize.

We sat down in the hangar's shade with the two Iranian kids, who had met three of their countrymen from another boat—they'd arrived a day before us but still hadn't finished registering. They were in good humor, particularly Firouz, an Iranian Kurd in his sixties, short and spry, with a salt-and-pepper mustache and bushy eyebrows.

"Don't worry, boys, we'll be in Athens in no time," Firouz said, making space on some pieces of cardboard for Omar and me. "We've got it all figured out."

It turned out they'd been sent on the boat by Hajji, too. "Well, you're all set then!" Firouz exclaimed. "He's got a guy who can get us off this island, easy. Twelve hundred euros each. Once we get into the main camp, we'll get a Greek SIM card and call him. If you've got the money, we'll all go to Athens together."

Lunch was a carton of rice and lentils with pita bread, our first meal since the kebabs in Izmir. Afterward I couldn't keep my eyes open. Omar and I both passed out on the hangar floor, and when we woke up the offices had reopened and we'd lost our place in line. It was dark by the time we got to the medical station. An Afghan woman and her teenage daughter were waiting outside, both with the same oblong, delicate faces. They'd come from the main camp, where they'd been stuck for several months.

"How is it?" I asked.

"Terrible, terrible," said the mother. "There are people from so many different countries here, that's part of the problem. There are a lot of Africans. But you know what? When we came over, our boat was full of blacks. And they were so kind to us families. They carried our bags down to the beach, and they fought off the Turkish coast guard." She giggled. "So I say, God bless the Africans, may their hands not ache."

Her daughter went in to see the psychologist alone.

"So you came straight from Afghanistan?" the woman asked. They'd been living as refugees in Iran.

"Yes, from Kabul."

"The kids these days all speak like you," she said teasingly. From my looks, she'd assumed I was Hazara, too. "*Sahist, sahist*, in that Kabuli accent. What ever happened to your Hazaragi?"

The exam was a hands-off affair. I sat in a chair while a Greek physician from Doctors of the World ran through his checklist. I had trouble with the Iranian translator's terms for things like "prescription," and he didn't seem to know Dari dialect, because he kept repeating himself angrily until I figured it out. Yes or no: Do you have any serious illnesses? Are you on any medications? Do you have any disabilities? Psychological problems? Have you been a victim of torture, abuse, or sexual assault?

The doctor told me to call the next person.

The mother and daughter were gone, but there was a line of glum-looking young men sitting on the bench outside the police container. I learned later that this was the nightly catch of people trying to sneak on the ferry.

When Omar and I returned to the hangar, we saw that the others had completed the procedure and been released into the main camp. But it was too late for us now, the cop said. We had to spend the night here.

We could hear music and drumming outside, and see crowds passing along the road. One kid came up and shouted: "Where are the Ethiopians?" He had two women with him and was holding what looked like a bottle of liquor. A cop shooed him away.

Another inmate, lean in a loose T-shirt that showed his tattoos, came over and offered to sell us cigarettes and SIM cards. He introduced himself in English as Abu Adam, from Palestine. I asked him what the situation was in the camp.

"People will cheat you for one euro here," he said, leaning close to the chain-link. "The first day I was in there, like you. I gave a guy outside ten euros to get me a SIM. I still haven't seen him. Don't trust anyone, OK? Don't even trust me right now while I'm telling you this."

He gave a sideways glance, and slipped back into the crowd of migrants. I walked to the hangar, exhausted.

It was getting cold again. A young American came over with a box full of fleece blankets, navy blue with SAMARITAN'S PURSE® emblazoned on them. Omar and I wrapped ourselves, lay down on pieces of cardboard, and were soon asleep.

14

When dawn came to the olive groves, the sparrows took wing. Down below, the sun warmed the tents strung between barbed wire. At the inner gate, Omar and I stood with our packs on our shoulders, holding the sleeping bags we'd just been given by an official as we were released into the main camp, who told us to get our tent from an NGO called EuroRelief, located up the hill.

We stared, shocked by the squalid scene in front of us. Lining the muddy gravel road were the charred frames of containers and UNHCR huts that had burned during the fire the week before. Around them, the orange-and-blue domes of two-person camping tents were crammed into every patch of open ground, with tarps lashed overhead for shade with rope and bits of wood. The stench of sewage hung in the air. As we followed the road uphill, we could see, through unzipped doorways, the inmates stirring from their sleeping bags, yawning as they pulled on shoes. Others trudged down past us, on their way to the toilets.

We found the EuroRelief trailer halfway up, across from the big tent where meals were served. Omar went up to a young woman wearing a Mennonite's bonnet and gown and explained in English that we'd just arrived. She took our registration papers and went to get her laptop.

As we were waiting, a group of inmates started vying for the other staffers' attention. They said they'd lost their shelter and clothes in the fire, but were told that the tents were only for newcomers like us, and that clothes distribution was only on such and such day, at a specific hour—

"You gave out shoes yesterday!" one man shouted.

The Mennonite returned with two Americans in shorts and polo shirts, one carrying a tent. "They'll find you a place and help set your tent up," she said. Omar and I followed the two volunteers down the road.

"The camp is full to bursting right now," one said apologetically. They had midwestern accents and the squeaky demeanors of college students. "We found a place down by the toilets, but it's not a great one."

"We would prefer somewhere near other Afghans," Omar said.

"Oh, right," she answered. "Well, we can check the upper level, then?"

"Are we allowed to do that?" her colleague asked.

She shrugged, and we turned back uphill. Across the road from the chow tent were two large, fenced-off housing blocks reserved for families. A narrow strip of gravel ran between their upper and lower terraces and ended at a fence. This blind alley was just wide enough for a row of the rectangular UNHCR huts, which had survived the fire here, but people had squeezed the little domed tents into the space that remained. At the end, we found an empty spot.

As the women fit together tent poles, the alley's denizens came out to watch.

"They're putting more tents in here?" one groused in Dari.

"Yes, but what can we do?"

Omar greeted them.

"Don't put your tent there," said a man on crutches, thin with gray hair.

"Why not?" Omar asked.

He pointed to a pipe coming out of the retaining wall.

"It's a drain. Dirty water comes out."

We picked up the tent, and moved it between two of the huts. Thanking the volunteers, Omar and I walked with them to the entrance of the alley.

"How come you speak such good English?" one of them asked.

"We were translators for the coalition forces in Kabul," Omar said.

Her colleague patted her pockets. "Oh, my God, I think I lost my phone."

They rushed back and started searching in the gravel. The gray-haired man reappeared.

"Did you lose something?" he asked them in Dari. He pulled an iPhone from his pocket and handed it to the relieved woman.

"If it had been anybody else, I would have kept it, by God," he said to us, after the Americans departed. His face lit up when he saw our cigarettes, so Omar offered him one. "I haven't smoked in days," he said, puffing lustily. "I'm completely out of money." He had been trapped here with his wife and kids for five months.

"Welcome to the prison," growled a tall youth with a bandage over his Adam's apple, stepping out of the hut beside us.

"Can we get out of the camp before our twenty-five days are up?" I asked.

"There's a hole in the fence in back. The camp isn't the problem, the island is the problem," the youth answered. He'd been here for four and a half months.

"You can't get off of this island," said a sonorous voice. We turned to see a man, with a thick walrus mustache, emerging from the other hut. He was missing one leg; using a crutch, he eased himself into a camping chair. "Only a single way exists: the trucks. Find a truck, one with salt, wood, or garbage that's going to Athens, and hide inside it. That's the only road of salvation."

"It's true," the gray-haired man added. "We've all tried, and we've all given up. You'll give up too, once you understand."

"But we know a smuggler who says he can get us documents for twelve hundred euros," said Omar.

The youth snorted in disgust and went back into his hut.

"Won't work," said the gray-haired man, dragging on the last of his cigarette.

"You'll never get off the island," said the other from his chair.

WE WERE HUNGRY AND OUR neighbors had warned us to line up early for lunch, so we stashed our bags in our tent and walked out of the alley, or *kucha* as its Afghan denizens called it. Our new home was right in the middle of the hill that Moria occupied. The camp's main entrance was at the bottom, next to the inner compound where we'd spent the night. From the gate, the gravel road ran up in a loop past the terraces and chow tent, and back down to a second fenced-off compound where the asylum interviews took place, next to the toilet blocks.

Built for two thousand people, by that point there were around five thousand crammed inside Moria, with hundreds more arriving each week. The tent sprawl was informally divided into neighborhoods by language and nationality; across the road from our *kucha* were Arabs, farther down it was mostly Africans, and at the back of the camp you had Pakistani and Kurdish blocs. The camp lacked adequate food and sanitation, and tensions among the inmates were high. The police were mostly concerned with protecting the inner compound where the officials worked. At night, they left the inmates to their own devices, even though there'd been rapes and robberies. Moria had been teetering on the edge of disaster for months.

The week before we arrived, a group of Arab and African inmates had tried to organize a hunger strike, but the Afghans had refused to participate. The trouble started at lunchtime, when families were served first. I heard two versions of what happened: one was that the hunger strikers had simply blocked the Afghan women from entering the meal tent; the Afghan version was that they were insulted and had the trays knocked from their hands. A brawl ensued, which escalated into a rock-throwing melee. The police and camp staff retreated into the inner compound, whose impressive defenses were to keep people out as much as in. Some rioters threw burning rags onto one another's huts, and the fire spread swiftly through the camp as the inmates, most of whom had nothing to do with the violence, fled into the fields. Darkness fell and the orange flames leaped high as their belongings burned. The fire department arrived in time to save the inner compound. Somehow, no one had died, but people were now living in flimsy tents in the mud.

When Omar and I arrived at the chow tent, there was still an hour until the men's turn for lunch but there was already a long line. The food was one of the main grievances at Moria; it was terrible, and there wasn't enough. The camp served boiled, starchy fare like macaroni or potatoes. If you got there before they ran out, there might be sides like an apple or pita bread, sometimes even beans or a bit of meat. There wasn't enough to go around, and that made people pushy. Normally, the police supervised the queue, but on our first day they hadn't shown up. There were some municipal workers there, but no one was listening to them. Shoving and shouting, people argued over whether they could save spots for friends, while those farther back whistled and jeered. The line in front of us kept getting longer.

Eventually, Omar and I gave up and stood to the side, next to one of the workers, a Greek with long, metal-dude hair. He lit our smokes for us.

"Is it always like this?" Omar asked him in English, watching the chaos.

"Sometimes."

The cops finally showed up, and started yelling at some miscreants to get out of the queue.

"What will the police do if they don't listen?" Omar asked.

"They will beat them," the attendant said, matter-of-factly. "But don't worry, you look like peaceful guys."

Having to constantly wait in line for something you desperately need and might not get can change a person for the worse. In the Australian detention camp on Papua New Guinea, asylum seekers spent their days lining up for meals, bathrooms, doctor's visits, phones, cigarettes, even malaria pills. *The queues have agency and they establish something: any person in the prison who behaves in a more despicable and brutish manner has a more comfortable lifestyle*, wrote the Iranian

Kurdish author Behrouz Boochani, who spent four years there. *We are a bunch of ordinary humans locked up simply for seeking refuge. In this context, the prison's greatest achievement might be the manipulation of feelings of hatred between one another.*

The chow tent in our camp had a dove with an olive branch on it; meals were served by Mensajeros de la Paz, a Spanish group founded by a Catholic priest. One of the stranger things about Moria back then was that it was largely run by Christian volunteers, many of them American, like the women who had decided where to put our tent. Their group, EuroRelief, was a Greek evangelical charity, little known before the crisis, that had arrived on Lesbos the previous summer, to serve tea on the beach. Then, eager to help the refugees they saw on TV, a torrent of foreign volunteers had descended on the Greek islands. The American megacharity Samaritan's Purse, run by the preacher Franklin Graham, partnered with EuroRelief, swelling their ranks. *He has a plan,* two women who joined from Minnesota wrote on their blog, *and maybe a tiny part of that plan is to bring Muslims to Him by allowing them to see His love through a cup of hot chai.*

Although by then there were around eighty NGOs working on Lesbos, so many volunteers were still coming that some groups charged fees to work with them. When the border was open, the NGOs had helped refugees travel onward to Europe; now they assisted them in captivity. After Moria and the other island camps became detention centers as a result of the EU-Turkey deal in March, the UNHCR and some of the other humanitarians, like Doctors Without Borders, pulled out in protest. The Greek government and the EU were caught flat-footed; they'd been relying on the nonprofits to provide basic services like food and health care. The whole hotspot system might have collapsed, had EuroRelief not decided to stay and take over the camp's vital functions. Within a few days, Samaritan's Purse and some other groups returned as well. Moria was saved.

THE EU'S HOTSPOT SYSTEM HAD two main components: First, incoming migrants were registered and fingerprinted, as Omar and I had been. Second, migrants who applied for asylum would be examined inside the camps. The only legal way to get to Athens was to pass your preliminary interview; then you'd be allowed onto the mainland for your final assessment. Those who failed and came from a country that had a *readmission agreement* with the EU, like Pakistan, could be deported from the island.

In the mornings, a crowd of inmates would gather at the fenced-off compound near the toilet blocks, where the asylum interviews were held. When I joined the people listening for their names to be called for interviews, I recognized, among those admitted, a few of the Syrians from our boat. Maybe the process would go faster than Omar and I had thought. "Are they calling Afghans today?" I asked the Afghan kid standing next to me.

"How long have you been here?"

"I just arrived."

He laughed. "I've been here two months and I still haven't gone for the first interview. They just started calling Afghans who arrived four months ago."

So our neighbors in the alley were right about being stuck here. I watched the people stretch their papers through the chain-link. The officials and their interpreters on the other side were from the EU's asylum agency which, like Frontex, was staffed by delegates from various member states. Though Moria's inmates were applying for asylum in Greece, their cases were heard by EU officials, who were in de facto control of the hotspot. The Greek government, which had almost been kicked out of the eurozone over its debt crisis, now risked being walled off from the rest of the Schengen zone if it didn't allow this extra-sovereign exception on its soil. *It's not mandatory, but in practice it's quite mandatory*, the Swedish interior minister had explained.

The EU officials had guards beside them, unarmed but wearing what looked like stab-proof vests, from the multinational G4S, whose contractors I'd also seen in Iraq and Afghanistan. The guards, working-class Greeks, were clad in shapeless uniforms while the officials wore trim oxfords and chinos. The EU interpreters dressed like their colleagues but had the faces of those outside the fence.

"When did you leave Kabul?" I asked the Afghan boy.

"A year ago." He smiled ruefully, and showed me the scars on his wrists, pales lines in his deep tan. "We were kidnapped by thieves in Iran."

When the officials were done calling names for the day, the disappointed crowd thinned out. I saw a sheet of paper, posted on the fence, which listed the twenty-eight countries of Moria's applicants: Afghanistan, Algeria, Bangladesh, Burkina Faso, Burundi, Cameroon, the Dominican Republic, Egypt, Eritrea, Gabon, Gambia, Ghana, Guinea, Haiti, India, Iraq, Iran, Lebanon, Libya, Mali, Morocco, Nepal, Niger, Nigeria, Pakistan, Palestine, Senegal, and Syria.

Some of our people from our boat would get to leave the camp faster than the others. The Syrians were given priority for asylum interviews, as well as for housing in the family camps. They were, in general, treated with more consideration by the police and aid workers. Even Pope Francis, when he took refugees from Lesbos on his plane back to the Vatican for asylum that April, brought only Syrians. Afghans were in the middle of the hierarchy; people from countries like Senegal or Pakistan were considered economic migrants unless they could prove otherwise. In Moria, the inmates had to compete for food, shelter, and medical care, but what they most wanted was to leave, and this pecking order further divided them. The Syrians complained of the others who'd crowded into Europe when the border had been opened for their sake. The Afghans were bitter that the Syrians got more sympathy, when their own war had lasted decades longer, but were quick to say that Pakistanis were not real refugees. An Eritrean told me how much he resented the West Africans, who weren't escaping a dictatorship like him. And the Pakistanis and Senegalese could retort that everyone had left Turkey for the same reason: a better life in Europe.

Prejudices based on color or creed had existed back home, but in the camp they were animated by a new logic, one which justified the way of the world. The migrants were learning to see themselves through Western eyes.

15

I followed Omar through the tents at the top of the hill, beneath laundry lines and past a group of Pakistanis playing cards. "Here it is!" he exclaimed. We'd found one of the exits cut in the perimeter fence by the inmates. The holes weren't a secret; everyone knew about them, including the police, but they served as pressure-release valves for the camp. Omar and I weren't supposed to leave Moria for our first twenty-five days, yet here we were, ducking under the jagged chain-link and stepping out into a grove of scraggly olive trees, careful to avoid the feces underfoot. The holes were so well established by then that there were food trucks parked nearby along the access road. We walked down to the main gate, where there were more trucks serving souvlaki and coffee. A collection of plastic tables was rigged with power strips so people could charge their phones, outlets being scarce inside.

It was a relief to stand outside the wire, but we were still trapped on Lesbos, even though we were now within the supposedly border-less Schengen zone. Tourists who came from Frankfurt or Amsterdam to spend their euros on the Greek islands didn't need passports, no more than Americans did for a trip to Miami. With an ID card, sand, sun, and ouzo were just an EasyJet flight away. But the boat people were not allowed to leave the islands, not without passing their first asylum interview; the police and Frontex had locked down the airport and ferry.

Firouz, the Iranian Kurd we'd met inside the inner compound the first day, had claimed we could get off the island with a smuggler. When Omar called him, Firouz said he'd already escaped the camp, and that we should come meet him in town. Omar and I waited at the stop by the main gate until a battered city bus pulled up; we paid our fare and sat down. A special service ran in a loop from Moria to the port; it took us inland through the arid hills past an army base, and then south toward the Bay of Gera, where we stopped at a smaller camp for families, and several women with strollers got on. Then we chugged uphill, past abandoned car dealerships, the bay glittering in the sun behind us. It was here, in the shallow lagoons of Lesbos, that Aristotle undertook the first systematic study of natural life, arguing that the soul was not separate from the body and that we were all connected in a *great chain of being.*

At the crest of the hill, we entered the outskirts of Mytilene, passing pharmacies with neon crosses, mechanics shops, betting parlors, then the town hall, arriving at the old fishing harbor, once the headquarters of the Ottoman navy, now with several Hellenic Coast Guard ships and UK Border Force cutters alongside. At the far end, behind the police dock where our Norwegian rescuers had taken us, were the quays for the massive ferries that went to Athens. Our final stop was the main square, which featured a statue of the bard of love, Sappho, whose poetry survives in fragments:

> *someone will remember us*
> *I say*
> *even in another time*

The plaza was ringed by cafés, and we spotted Firouz sitting outside one with his two traveling companions: Arash, heavyset with a shock of white hair, the cafeteria manager at an oil company in Tehran, and Reza, in his twenties with a gelled flip, who'd worked there. All three were trying to get to Germany.

"You're staying in the camp?" Firouz asked. He made a face. "It's disgusting there." His wife, who was already in Germany, had connected him with a new smuggler, an Egyptian woman, who'd arranged a hotel, for which you normally needed papers, at forty euros a night. The smuggler was charging the same as Hajji had quoted, 1,200 euros for fake documents and a flight to Athens, and she claimed she would bribe an official at the airport to get them through. She wanted four hundred euros in advance, but Firouz wasn't worried. "She's better than Hajji. It's a hundred percent," he said, tapping ash off his Winston. He and his friends were leaving as soon as the woman was ready, hopefully by the end of the week. He promised to introduce us to her. "Look, if you're interested, try to find some other clients in the camp, and she'll give you a discount," he said.

"What do you think?" I asked Omar once they'd left.

"I think we should wait and see what happens," he said.

We looked around the square and could see a few migrants hanging around. We'd heard the police didn't bother you in town and neither of us felt like going back to the camp, so we walked toward the ferry terminal, where someone had painted, on the long concrete breakwater, SMASH THE BORDERS. Last summer, thousands of refugees had been camped around the terminal, waiting for their turn to board the ferry. Some shuttered travel agencies still had posters up in Persian and Arabic offering bus tickets from Athens to the Macedonian border.

On the other side of the perimeter fence, we found a concrete swimming platform, pebbly and oil-stained, with wooden changing booths and a shower. A couple of Afghan kids were lying in the sun there, and waved to us. I looked down at the cloudy blue water. My skin was crawling with grime from the camp, so I stripped to my underwear and dove in. The two boys jumped in after me. But Omar, always first into a pool or lake back in Afghanistan, stood in his boxers at the edge of the pier.

"Come on, what's the matter?" I called, treading water.

"It's salt water?"

"Yeah. It's the sea."

"I've never swum in salt water. It won't burn me?"

I laughed and kicked back, letting the water run over my nose and lips. After a minute, Omar jumped in, and came up smiling. The four of us swam out and bobbed there. If you lay on your back and rolled to one side, the ferry's red smokestack loomed. To the other, the sky and sea merged into a pale sphere.

Afterward, we rinsed off in the open-air shower then lay on the concrete to dry out. You could tell the kids had been on the island for a while by how skinny and tanned their bodies were. Ali, seventeen years old with frosted blond tips, said he'd been here for six months. He'd landed only a few days after the islands closed in March and back then you could still buy a counterfeit asylum card—the kind they gave you once you passed your first interview—and get on the ferry to Athens.

"They cost a hundred and fifty euros," he said. "But they said our interviews would happen soon, so I thought, why pay? Then papers went up to three fifty, then six hundred, then a thousand euros, and now they don't work at all." The cops kept a tight watch on the ferry. If you were lucky enough to get sent to Athens, the UN had to send someone down to the port to escort you on.

"The police don't let you inside the port, but they don't bother you here," Ali told us. He pointed to some blankets and bags stashed by the fence. "Some of the boys sleep here."

We climbed up the hill behind the platform, the sun warming our skin. At the top was a bronze statue in classical robes atop a bas-relief plinth, her breasts bared and her right arm holding aloft the torch of liberty. She gazed defiantly at the Turkish coast, visible on a clear day like today. It was one of several monuments on the island to the Exchange of Populations in 1922, when Mytilene's streets were crowded with sick and homeless refugees. Many of them settled here on the island, where their villages were still referred to as *prosfygika*: refugee towns. Some of those expelled from Turkey didn't even speak Greek; under the Treaty of Lausanne, people were forcibly deported on the basis of religion alone, formalizing the alchemy which transformed Ottoman Muslims and Christians into Turks and Greeks. This ethnic cleansing was supervised, in its final stage, by the League of Nations; at the conference in Lausanne, the polar explorer Fridtjof Nansen, who had become high commissioner for refugees, proposed a complete swap that would *unmix the populations of the Near East.* That year he was awarded the Nobel Peace Prize.

We passed the rest of the day at the harbor, and took the last bus back to Moria at ten. Some of the passengers were visibly drunk. Alcohol was banned inside the camp, though you could bring in whatever you wanted through the holes. "I could use a drink myself," I told Omar wistfully as our bus pulled up to the gate's floodlights. "We should have gotten something."

"Wait here a minute, I have an idea," Omar said, and strode off. When he returned, he was carrying a brown paper bag and motioned for me to follow. We walked farther up the road, into darkness. He showed me two tall cans of Alfa beer.

"Where did you get that?" I asked.

"From the trucks."

We walked and found a flat nook beneath some olive trees, where we cracked the cans and sat looking out on the moonlit hills. Behind us, you could hear the rattle of generators from the camp. Had we made a big mistake by coming to the islands? The situation was much worse than I'd imagined. Glancing at Omar, I wondered if he thought I was crazy to stay here with him.

"It's-a cold but it make-a you warm, brudder," Omar said, slurping. I snorted my beer—he'd taken to speaking garbled English in parody of my own phony Afghan accent, the one I was using with aid workers.

"Cheers, brudder."

"Cheers."

To get away from the camp, Omar and I started taking the bus into town every morning, leaving after breakfast and staying late. If we were lucky, we got lunch or dinner from one of the charity groups that came to Sappho Square; other days we'd spend a few bucks on a pita sandwich. We had only five hundred euros that I'd sewn into my pants, and we had to make them last. We'd planned for his mother to send us money from Istanbul, but there was no *hawala* on Lesbos, and neither of us had any documents, so we couldn't use the Western Union here. Later, as our stomachs shrank, it became difficult to eat a whole restaurant meal in one sitting anyway.

The swimming platform at the port had been claimed by a band of Afghans from our own alley. They'd set up a tent and some blankets under a couple of trees, and we passed time there playing cards and scheming ways off the island. The guys were a mixed group; although Afghan migrants typically traveled through networks organized by hometown and ethnicity, and though there were often tensions between Sunnis and Shias, or Persian speakers and Pashtuns, when faced with the violent international rivalries of the camp, the Afghans stuck together.

Some of the *kucha* boys were wanted by the police for their role in the riot and subsequent fire, and visited the camp only late at night through the holes.

"We were outnumbered and they beat us up, so we had to get revenge," one kid told us. "We soaked some scarves in fuel, lit them, and threw them on the Africans' tents. Now they know not to mess with us."

One man, who hid his scar-nicked crewcut under a bucket cap, had already been inside the camp jail on knife charges, awaiting transfer to the mainland. When the fire started, the guards fled, and he and the other inmates broke down the gate and escaped. He came from the slums on the west side of Kabul, and in the camp he'd found a similar scrabble for survival. One day, after he drove off some Pakistanis with blows and curses, I asked why he was always getting into fights.

"You just arrived," he sneered. "If you'd been here before, you'd have seen how many beatings Afghans took from Pakistanis. It's calm in the camp now, huh? That's because they're afraid of us after the fire. You can line up easy for food, because we fought."

For the most part, though, the guys didn't bother the other refugees who came to bathe or to fish with lines baited with dough. The other end of the platform was, by tacit arrangement, used by elderly locals. There were not many other visitors, but a couple of times I saw a white hatchback cruise by, a pale arm on its windowsill.

"*Ena! Ella!*" the guys catcalled.

"That *mordegow* comes every day," the guy in the bucket cap said.

"Why?" I asked.

Our one-legged neighbor snorted. "Give him your ass and he'll give you fifty euros, that's why."

"So it's true!" exclaimed Omar. Some of the boys looked sheepish.

Every evening, the ferry departed on a twelve-hour journey to Piraeus, the port of Athens. As the afternoon wore on, the migrants gathered by the fence to ogle her stern ramp. Though you needed to show valid travel documents to buy a ticket, you could get one on the black market at an inflated price, but the police waited by the gangway and would check your papers if they suspected you were a migrant. Afghans and Syrians, with their lighter skin and more Caucasian features, had a better chance of passing as a local or tourist than Pakistanis or Moroccans, who in turn had it much easier than Eritreans or Senegalese. The boys kept encouraging Omar to try.

"You look like a foreigner," Ali, the seventeen-year-old, told him. Ali was Hazara, and looked quite Asian. "And you speak good English."

"What about me?" I asked.

He laughed and pulled the corners of his eyelids. "Sorry brother, you're like me."

If you couldn't pass, that left the trucks.

Much of the ferry's cavernous vehicle decks were taken up by tractor-trailers hauling freight to and from the islands. If you were feeling reckless, you could crawl beneath a trailer and cling on above the axle, like Odysseus escaping the Cyclops under the belly of a sheep. Many tried, especially nimbler kids like Ali. But the trucks were searched as they boarded, and it was easy to spot someone underneath them. It was better to hide inside the cargo, but the parking lot at the port was too well guarded. You had to find your truck elsewhere on the island, and break in. Lumber yards, quarries, and recycling depots were good bets, since the trucks there were usually headed to the mainland. You might have to wait inside the trailer for a day or two, before it started moving. It could take many tries before you made it through to Athens, and each time you risked being crushed or suffocated in a steel box.

Whereas in Iran and Turkey, the Afghans had been passive *travelers* who belonged to smugglers, here on Lesbos they were players in what they called the *game*, using the English word. We talked game all day, about where to find trucks, or whether it was better to buy a roller bag or backpack to pass as European. When evening came, those who weren't playing went to the fence to watch. Boarding started at six. First the ferry conductors turned up, the ones with ID cards around their necks. They checked tickets. A Greek cop in uniform joined them, another manned the gate, and others walked the perimeter, yelling at us if we crowded the fence. The state was Argus-eyed; a buff dude with a bulge under his Hawaiian shirt was lined up with his fellow passengers. The Dutch prison van arrived, disgorging large guards in reflective vests. They returned captives to Moria. Last came the Hellenic Coast Guard special forces, the commandos, as the migrants called them for their camo fatigues with the cuffs rolled over their biceps and boots. They put on rubber gloves and took out long steel flashlights.

A rumble went down the column of trucks as the drivers crept forward to the ramp, where the commandos probed underneath and pulled the cargo doors. The passengers filed past the conductors, not many tourists this time of year but plenty of NGO and camp workers. By now it was getting dark.

The migrants kept up a commentary at the fence: who was trying tonight, who'd bought a bag and got a haircut, who'd been in a truck since yesterday.

"Where are you? Inside?" the guy next to me whispered into his phone. "Just stay quiet and hang onto something, you'll pass like that."

We rooted for the players. That might be us tomorrow.

"There goes Jawad," someone hissed, "he's wearing shorts." A kid towing a roller bag darted from the shadows of the parking lot, blending deftly with passengers walking up the ramp.

"He had a ticket, he made it—oh."

Two cops led him by the elbow down the gangplank. Even on-board the ferry, you weren't safe—an Iranian had made it to the top deck before he was spotted by an undercover cop.

Another groan. The commandos were yanking two kids out from under a truck. They usually weren't rough, maybe a swat on the head if you were slow, but otherwise you went into the Dutch van. If it was already full, they might just tell you to scram—*fyge malaka*. As long as you weren't carrying fake documents, the worst you'd get was a night or two in the camp jail. The cops could net twenty players or more in a night. But the real question was how many made it through. The receding ferry's lights glittered in the humid night, as she gave a long blast of her horn.

THE ISLAND IS A PRISON, wrote Christos Ikonomou in one of his short stories about the Greek economic crisis, *and the sea is the bars.*

One night we heard Rage Against the Machine playing on a ste-reo in Sappho Square; a few hundred people had gathered for a rally against what they termed *fascism*, the xenophobic parties from Athens that were gaining popularity on the islands. The group of mostly young people marched in a loop through the town center, past the ice cream and ouzo stands, past the sunglasses boutiques, chanting in Greek and English: "Say it loud, say it clear, refugees are welcome here!"

Omar and I watched from a distance; afterward, he stared, aghast, at some graffiti from counterdemonstrators: FUCK ISLAM.

For the compassion they'd shown to the boat people the previous summer, the Greek islanders had been nominated for that year's No-bel Peace Prize. *We welcomed refugees because we're descended from refugees, too,* an elderly woman explained to the press. The previous September, eleven days after Alan Kurdi drowned, the mayor of Les-bos published a declaration along with his counterparts from the *cities of refuge* of Paris, Barcelona, and Lampedusa. *If we continue to build walls, close borders, and delegate dirty work to other states so that they act as gendarmes of our frontiers, what message are we sending to the world?* the mayors wrote. *We, the cities of Europe, want to welcome refugees.*

There was a bounty of awards that year for work done on Lesbos: the World Press Photo Award, the Pulitzer, the Olof Palme Prize, the Nansen Refugee Award from the UNCHR. The Council of Europe gave the Raoul Wallenberg Prize *for extraordinary humanitarian achievements* to a local solidarity group, Agkalia. *You know, after the prizes, they've invited us pretty much everywhere, in Sofia, in Vienna, you name it,* Giorgos Tyrikos-Ergas, a member of Agkalia, would later recall. *Whenever we start talking about the roots of things, about war and about how the Western world keeps pillaging other countries, the discussion is, "Ah, you're so right, but these things are so complicated. Let's focus on issues of how we should install refugees, relocations."*

The Nobel was going to be announced on October 7—we even heard about it in Moria. Online, the betting shops had the Greeks as favorites, 13 to 8. The papers ran profiles of the two Lesbos residents selected as symbolic nominees, a grandmother and a fisherman.

The winner was the president of Colombia, for negotiating a peace treaty with guerrillas.

As the world's attention moved on, the islanders were left hosting camps for Europe's unwanted. The border had closed, but the boats were still coming. And the mood on the islands was changing. The people in their houses, as John Steinbeck once wrote, *felt pity at first, and then distaste, and finally hatred for the migrant people.* Individual acts were tallied against the common account; the refugees were blamed for thefts, for the coils of shit by the port, for sexist abuse hurled at policewomen. And the islanders were suffering, too. Tourists hardly visited the hotspot islands anymore, and Greece was still deep in economic crisis. There were brawls between locals and migrants; the Chios camp was firebombed. On Lesbos that fall, some parents chained shut the gates of a school, to keep refugee children out.

————

WE WERE SITTING ON THE bus one morning when Omar stiffened and nudged me. An Arab girl, probably Syrian, was getting down with an infant in a stroller; she had Laila's faint eyebrows, her skin pale as cream under her dark veil. Omar whispered hoarsely: "She looks like her. Exactly like her."

We rode that bus each day except on Sundays, when it didn't leave until the afternoon, so we walked the five miles into town instead. On the way we passed groups of African migrants, escorted by camp Christians, coming from church neatly dressed in collared shirts and dresses, the women with hats like tropical birds. At the intersection, matronly Greeks stood handing out Jehovah's Witness tracts in Persian, Urdu, French, and English. By now, a glance was all they needed to come out with the right pamphlet.

My Persian tract had flames on the cover, but inside it said that because God was good, he would never torture his children in hell.

"You know, Muslims believe that too," Omar told me as we continued down the road. "They believe that no one will be left in hell forever, because God is merciful."

"That's good news for me, then."

"But you'll have to pay for your sins first." He laughed. "Don't worry, brother, I'll sponsor you to come to heaven."

"I'll go by the smuggler's road."

"There's no smuggler's road over there."

"There's always a smuggler's road."

FIROUZ CALLED AND TOLD US to meet him at the café in Sappho Square. Seated with him was the Egyptian, a squat, henna-haired woman, her eyes hidden behind Gucci sunglasses. She looked Omar and me over briefly, as Firouz translated for her in Arabic, then said she'd send us next.

Later, we met up with the three Iranians in the park behind the theater. They'd shaved and put on new dress shirts, and bought three of the same roller suitcases, the cheapest one. They showed us the papers the smuggler had given them; Reza, the young guy, had a French passport, and the likeness wasn't great, which worried him. When I read the name out loud, he thanked me—he hadn't known how to pronounce it.

"Don't worry, it's a hundred percent," Firouz told him. "They won't ask you any questions. She has a guy at the airport they're paying off."

Firouz and Reza were going that same evening. Arash's flight was the next day. They were nervous because the Greek courts sometimes gave six-month sentences for using fake documents, which meant prison on the mainland.

Back at camp, we saw what must have been their jet fly overhead, its fuselage catching the last light. "Imagine being on it," Omar said wistfully.

In the morning, he called Firouz, who recounted how things had gone. The smuggler dropped them off at the airport. They went in together and made it through check-in, but when the security guard ran Reza's passport through the machine, she frowned and left, returning with a cop, who took them to a back room. Another officer came and spoke to Reza in French. Reza just shrugged, but when the handcuffs went on, he started crying.

"It's OK," the cop soothed.

He took away their fake papers, frisked them, and then drove them back to Sappho Square, where he took off their cuffs and told them to get lost. That was it. They were lucky—whether or not migrants faced charges seemed to depend on how busy the cops and courts were that day.

The Egyptian woman wasn't answering her phone. The Iranians were out eight hundred euros, plus the two hundred they'd spent on the hotel. Firouz was so angry he'd gone down to the police station to complain. They had just laughed at him.

When Omar hung up and told me the story, I could tell he was taking it hard. "That means I'm stuck here," he said. "It could be six months."

"We're not gonna stay here," I answered. "We'll find a way off."

He smiled without saying anything. We both knew I could make a phone call to get my passport back and leave, any time I wanted.

WHENEVER WE RAN INTO OUR boatmates in camp, we looked at each other in recognition, sharing thoughts if not language. We were conjoined by the water passage. One day a Syrian boy showed us a video that he'd taken on his phone of the moment of our rescue. The frame jerks from the blinding glare of the Norwegian cutter to him and his friends, wide-eyed and jubilant, the audio raw with adrenaline. I am a dark blur in the background. *His power was to have existed amongst them as a native speaker, as it were,* Edward Said wrote, *and also as a secret writer.*

The wall was inside of us, too. When the bus climbed above the bay in the morning I gazed from face to tired face until my eyes welled up so I had to look down at the scuffed row of sneakers, and it felt like shining a light on a mirror, what loving God must feel like, but in my heart I said it was just pity.

What did it matter, what I felt? I wished I was a camera eye. But at night I crawled into our tent and lay in my bag, slowed my breath, and listened for layers: Omar beside me, our neighbor's laugh, the clink of spoons and the sound of water, then a more distant circle until, if I arrived at perfect stillness, I'd hear Moria's five thousand souls.

16

In his bleak portrait of British poverty in the 1930s, *The Road to Wigan Pier*, George Orwell traveled to the mines and slums of the industrial north. *It is a kind of duty to see and smell such places now and again, especially smell them, lest you should forget that they exist,* he wrote. Rather than his readers' sympathy, Orwell was appealing to their intellectual honesty. He wanted to tell them something about themselves. The coal mine, with its brutal conditions, was *the absolutely necessary counterpart of our world above*, where everything *from eating an ice to crossing the Atlantic* was dependent on this fossil fuel. The miserable cottages that Orwell visited were linked by an invisible chain to his readers' comfortable living rooms. Orwell was a committed socialist at the time, but his argument was distinct from any political stance, radical or not. *You cannot disregard them if you accept the civilization that produced them.*

THE WEATHER HAD BEEN SUNNY since our arrival on Lesbos but one afternoon, during our second week, it clouded over and started drizzling. By evening, a downpour set in and washed all night, in waves, against the thin fabric of our tent while Omar and I lay awake feeling the water soak through our sleeping bags.

In the morning, we saw the camp was churned into a bog. Some

people's tents had collapsed in the rain, while others had found themselves in the path of racing torrents. My stomach wasn't feeling good,
so I went down to the bathrooms, past the inmates already lining up
for breakfast. The smell grew stronger as I descended: rotting trash,
unwashed bodies, and, most of all, shit, redolent of last night's lentils
and onions. The lower parts of the camp were flooded ankle-deep with
foul puddles. I got in line outside the men's stalls and lit a cigarette. As
I inhaled, I started coughing and my head throbbed; in truth, I'd never
been much of a smoker but I'd adopted Omar's habits on this trip and
he could go through a couple of packs in a day. And cigarettes did help
with the stench and hunger.

Once inside the bathroom, even if you breathed through your
mouth, you could still taste the miasma. A single bulb lit floors and
walls coated in grime. On the left were sinks and six mirrors, four of
them shattered. To the right, a row of stalls was crudely fashioned from
sheet metal. My shoes squished against the concrete as I entered one of
the squat toilets. There was no paper but there was a rubber hose lying
in the muck. I latched the door, then unbuckled my pants, carefully
bunching the material between my knees so that it wouldn't touch the
floor as I squatted. My thighs were already straining as I lowered myself toward the hole.

EFFECTIVE BORDER PROTECTION IS NOT *for the squeamish*, the former
Australian prime minister Tony Abbott had told a gathering of politicians at Lobkowicz Palace in Prague the previous month. *You have to
match the conviction of those demanding entry with the greater conviction that you have a right to say no.*

Abbott was extolling his country's Pacific Solution, where asylum
seekers were intercepted at sea by the navy and transferred to remote
islands for offshore processing, where they would wait for years inside
camps, bitten by mosquitoes that carried dengue and Zika viruses.
Some would eventually be sent to a third country, but none, regardless
of the merits of their asylum claim, were allowed to settle in Australia.
Popular with a majority of Australians, the policy had been effective;

twenty thousand people had arrived in 2013; zero in 2015. Mass incarceration was one answer to mass migration.

Europe's challenges are on a larger scale and the geography is different, Abbott said, *but with the right will and organization there is no reason why there could not be similar success.*

I STOOD OUTSIDE THE TOILET blocks, trembling, and lit another smoke. The line was longer now, and wound its way through dry patches in the sewage. I counted seven men's stalls in the lower block, and another twelve uphill; a third block was out of order—the camp's water often stopped running entirely. Nineteen toilets for several thousand men, along with some porta potties. The women had their own toilets but they were unsafe after dark, as were the fields outside the fence.

I walked back uphill, my cough joining the chorus around me. My throat was raw; it felt like I was catching something. Some kind of respiratory ailment was going around—maybe influenza. The children, especially, had swollen eyes and mouths. There'd been outbreaks of scabies and varicella in Moria; our bodies were covered in insect bites and rashes.

MASS INTERNATIONAL MIGRATION IS A *response to extreme global inequality*, wrote the economist Paul Collier, explaining that the twenty-first century will see even more movement by people as technology makes it easier to see how things are elsewhere, and then travel there. *This is why migration controls, far from being an embarrassing vestige of nationalism and racism, are going to become increasingly important tools of social policy in all high-income societies.*

If borders are the necessary counterpart of an unequal world, some like Collier, who was knighted for his work on Africa, believe this gap will inevitably close thanks to the growth generated by global capitalism. *Mass migration is therefore not a permanent feature of globalization*, Collier concluded. *Quite the contrary, it is a temporary response to an ugly phase in which prosperity has not yet globalized.*

Nations like China have indeed been converging with the West but others have fallen further behind, particularly in Africa, which is likely to hold a quarter of the world's population by midcentury. Looking at the problem in an absolute sense, since only a tiny fraction of newly generated wealth goes to the poor, ending the kind of poverty that drives mass migration will, in the absence of radical redistribution, require us to increase the overall size of the global economy many times over—and along with it, our consumption of resources and production of waste such as greenhouse gases.

A warming planet means rising sea levels. One of the first countries to be submerged will likely be the Maldives, a low-lying archipelago in the Indian Ocean. The former president, Mohamed Nasheed, has explained that if wealthy nations did not drastically reduce their carbon emissions then Maldivians would have to flee, and people in the West would be faced with a choice: *When we show up on your shores in our boats, you can let us in. Or when we show up on your shores in our boats, you can shoot us. You pick.*

The border camps of the rich world will offer a third option.

THE ISLAMIC FESTIVAL OF ASHURA took place while we were in Moria, and in the housing block above our alley the Shia residents held an *azadari* ceremony in the courtyard. It was open to all, and some of the Christian volunteers came to watch the ritual, which mourns the massacre of Imam Hussein and his followers at the battle of Karbala in the year 680. That night I stood among the onlookers while perhaps a hundred men chanted as they raised one arm and struck their chest with the other, in alternating blows, the red palm prints appearing on their bare skin, inscribing a collective trauma.

Moria held the fragments of a former union, of homelands and languages left behind, and despite the brutal chaos of the camp, people wove themselves together here once more. The Afghan families living in the container next door had relatives and friends among the *kucha* dwellers, and at mealtimes they handed their extra rations to us through the fence so we wouldn't have to line up. Their children often wandered alone through the alley, watched over and unafraid. There were two little Hazara girls who liked to play on the embankment above us; with their matching black bangs they could have been Japanese. I could hear them from my tent:

"Salaam, teacher."

"Salaam. Everyone has ten days of holiday because of Eid."

"Really? I have a holiday too?"

Hundreds more refugees were arriving each week on Lesbos. You could hardly walk through our alley anymore, it was so crowded. Tents crept up the hillside, like the slums in Kabul. The boats landed at night and in the morning we saw the new arrivals in the reception area, sitting where we'd sat, damp and exhausted. When they got out, they always had the same question: "How do we get to Athens?"

Omar was coming back from breakfast when two young Afghans asked him for directions to the chow hall. Zulmay was twenty, tall and gangly; Raja, his cousin, was eleven and had the pouty looks of a tween singer. Five days earlier, they'd boarded a speedboat in Turkey at dawn with sixteen others, their little wooden shell so overloaded that its gunwales barely cleared the waves. Lucky for them, the sea was flat and calm that day.

Originally from Mazar-e-Sharif in northern Afghanistan, they were here in Moria on their own. Raja's mother—Zulmay's maternal aunt—had entrusted her son to his older cousin and charged them with making it to Germany, where Raja's elder brothers had already sought asylum. Because the boy wasn't with his parents, the camp officials wanted to treat him like an unaccompanied minor, meaning he'd be held in a locked dormitory within the inner compound. When Zulmay insisted that he was his cousin's rightful guardian, the authorities had kept them in the reception area for four days, trying to pressure Zulmay into giving the boy up. A young female social worker sat and held Raja's hand, serenading him, sotto voce, to come with her to the minors' section. There were other children there, she said, and besides, they were preparing new homes for them in Athens. She promised he'd be sent as soon as they were ready.

"And what did you say?" asked Zulmay.

Raja crossed his arms and pursed his lips brattily: "No!"

Omar and I laughed. Of course the boy didn't know about the suicide attempts that had taken place in the minors' section, which was, in essence, a prison; he just didn't want to be separated from his cousin. In the end, the officials relented and assigned Zulmay as the boy's guardian. Their tent had been in a place up the hill in a Pakistani neighborhood, but we helped them move to a spot near us in the *kucha*.

Zulmay and Raja asked to tag along with us when we went into town, so we all caught the bus and then walked from Sappho Square to the swimming platform by the ferry. The weather had cleared off again and the statue of Lady Liberty shone above us. The gang from the alley was absent and, for once, Greeks outnumbered refugees. We watched a group of elderly swimmers as they limbered up, donned caps and goggles, and jumped in, their bodies turning nimble in the water.

We went in too, except for Raja, who couldn't swim, even though his family had a place near the bank of the Amu Darya, on the border with Uzbekistan. His mother had been too protective of her youngest boy.

"Don't worry, I'll teach you," Omar told him.

As we sat drying off afterward, an owlish older man arrived on a bicycle. He greeted the other Greeks and then, observing the nearly empty platform, came up to Zulmay, who was wringing out the shirt he'd rinsed in the shower.

"This is not a laundry!" the man shouted in English.

Zulmay blushed and stammered: "I'm really, really sorry."

"You take all the space and we can't even swim here! You should go to that side."

Zulmay kept apologizing, and the Greek calmed down a little.

Omar came to his rescue: "I'm sorry, sir. We just came here because we have no place else to go."

"I know that if I went to your country, I wouldn't be allowed to walk around like this."

"Actually, sir, in my country, people respect the foreigners. We have no problem."

"No, I wouldn't be allowed because I'm a Christian. I'm an infidel. I don't mean you, but you have some crazy religious people there."

Omar was at a loss for words.

"Now we can't come swim here. I'm afraid. I don't know your intentions. There's been some incidents and people are afraid. You use a lot of water here," he said, pointing to the showers. "We pay for this. We pay taxes. It's an economic crisis here. I'm an educated person, and I've been unemployed for six years."

The man glanced over his shoulder at the other Greeks, who were watching.

"It's not your fault. There's someone behind you who's creating these problems." He gestured east, across the channel. "The Turks are making good money from you. It's human trafficking, and it's two hundred percent illegal."

He sighed and looked at Raja and me.

"Tell them it's nothing personal. I don't know if they understand. We just want a little respect. We're in the same situation, actually. We're both stuck in the middle of something much bigger than us."

Later, when it was my turn for the shower, the man came over and watched me lather myself.

"You're not allowed to use soap here," he said. He was wearing a Knights of Rhodes T-shirt.

I looked at him blankly. "No English."

He shook his head and rode off on his bike.

At sunset, we walked to the old castle above the port, where a group called No Border Kitchen usually distributed food out of their van, paper cups of rice and chickpeas spiced with cumin and cloves. Omar had been impressed on our first encounter, polishing off two helpings of the masala mix. "Much better than camp food," he declared. "Maybe tomorrow they'll have chicken."

"Omar, they're vegans," I said.

"What does that mean?"

"They only eat vegetables."

He shook his head and laughed. Henceforth, he christened the group No Border Chicken. A tattooed and dreadlocked crew of mostly

German activists, they were the only ones we saw minister to the men living in the woods below the castle, Pakistanis and Arabs who'd been rejected and slated for deportation. The other day, there'd been a scuffle between the two groups in line, but this time, when we brought Zulmay and Raja, everyone was calm and there was enough food for seconds for anyone who wanted. Omar had thirds.

It was a mild evening so we decided to walk back to camp. In the fading pastel light, we clambered beneath the castle walls and descended to the old harbor past the tourists sitting in fish tavernas. It was dark by the time we reached the highway. Ahead, there were other groups walking back to Moria, their shadows rising and falling with passing headlights.

Zulmay explained that he'd been studying medicine in Mazar, but when the border opened he'd left for Europe, nearly a year ago. He'd made it to Turkey and tried to cross but his boat had been intercepted twice. When the border closed, he gave up and went back to Istanbul, where he found a job in a restaurant. Then Raja and his mother had flown in from Afghanistan. She came for treatment for her diabetes, and had decided to send her little boy to his two older brothers, who were already in Germany. Her husband worked in customs at the Uzbek border, a lucrative position, and she offered to pay for Zulmay's trip if he took Raja along. So our new friends had money for a smuggler to get off the island. Omar explained how hard it was to leave legally.

"The Syrians and Iraqis are first in line for asylum, and everyone else has to wait," Omar said. "There's no justice here. All of Europe's talk about human rights is just bullshit."

"If only there was some sort of bridge or tunnel off this island," said Zulmay.

"The only way is by boat or plane." Omar said he wasn't crazy enough to hide under a truck, but he was thinking of paying a smuggler for fake papers, despite the risk of jail.

"We're willing to try anything," Zulmay said. "Raja, be careful, don't walk on the road."

I told the boy he could walk next to me.

"How's it going?" I asked.

"Not bad, but I don't think a spoiled kid could handle living in the camp," Raja said. "When I first came, and people said, 'Welcome to Greece's prison,' I felt like crying. But now I'm getting used to it."

Raja, I learned, was only afraid of three things: dogs, cats, and chickens. And his mother did not let him swim in the Amu Darya. But he'd driven a car a hundred kilometers an hour on the highway north of Mazar. What was the fastest that I'd gone?

"Probably about that fast," I said.

"What kind of car did you have in Kabul?"

"I didn't have a car. Omar drove us everywhere."

The boy was silent a moment. "When we get to Germany, we're gonna have a good time, right? We'll go sightseeing together, and eat big dinners. Do they have Afghan food there? You can meet my brothers and the rest of my family."

We followed the road inland through the village. In the distance, we could see the camp lights burning on the hill. A young German shepherd ran snarling to the fence as we passed. Other houses were deserted, perhaps summer homes, though some of the old stone mansions from the Ottoman times were little more than ruins.

"That house looks like it has djinns," said Raja. "Would you go inside?"

"I would," said Omar. "I once slept in a graveyard."

Zulmay played music on his phone's speaker, love songs by Ahmad Zahir and Nasrat Parsa, then a classic by Farhad Darya that we sang along to, raising our voices as we walked:

From this exile, from this wandering,
From this loneliness, from this prison,
To Kabul jan, to Kabul jan, to Kabul jan, salam!

ON SATURDAY, WE FOUND SOME activists in Sappho Square handing out flyers in English and Persian, accusing the EU and Afghan government of buying and selling refugees. The previous winter, while Omar and I were still in Kabul, the German interior minister had visited, and held a televised press conference with his Afghan counterpart, both men nearly obscured by a heap of roses, lilacs, and crocuses at their table. Germany, Thomas de Maizière had claimed, was willing to remain and help Afghanistan but Afghans had to stay in their own country—too many were coming to Europe for economic reasons. *That is understandable from a human point of view, but it doesn't give them the right to protection*, he said. Now President Ghani's government had at last, after much arm-twisting, signed the Joint Way Forward, agreeing to take back Afghans deported from Europe.

Later that evening, as the restaurants around the square grew lively with people and music, the No Border kids showed up with some Afghan families and started chanting slogans against the new agreement:

"Open the border!"

"Moria is a prison!"

"Afghanistan is not safe!"

The four of us stood by the statue of Sappho and watched. "What we need to do is get the whole camp together and protest," said Zulmay. He thought for a minute, then sighed. "No, they'll never let us leave. This is the best way to keep people from coming."

Word of the awful conditions on the islands had already caused migrants to shift to other routes, like Greece's land border with Turkey, but for us it was too late. We walked to the park behind the theater, where migrants hung out, and we bought a liter and a half of wine for three euros at a kiosk, plus a Coke for Raja. As we passed the bottle around, Omar, who'd been glum lately, grew animated and started recounting his exploits in Kabul, the big salaries he'd once earned from USAID, the times when he'd been broke, and the girl for whom he'd go around the world and back.

"Life is two days, brothers," he said, sighing. "The day you're born, and the day you die."

An Eritrean man staggered over, a beer in his hand.

"What kind of man is your president?" he shouted in English. "He's selling you!"

"You're right!" Omar leaped up and shook his hand.

Everyone took turns cursing President Ghani, who'd told the BBC that he had *no sympathy* for the migrants who wanted *to leave under the slightest pressure.*

"*Be namus!*"

"His kids live in America, and he wants us to go back to Afghanistan," groused Zulmay.

After we finished the first bottle, we got another. It was going down smooth. We made it onto the last bus back to camp, where I started speaking French with a startled Cameroonian, explaining I'd learned from the soldiers in Kabul. Zulmay was leaning on a strap, watching me. He was the only one who ever questioned my story to my face. We'd been walking together alone when he asked, apologetically, why my accent sounded a bit like a foreigner's. I covered the lie with another lie I'd prepared: the truth was that I'd grown up in Malaysia, I said, where my Afghan father had found a job, but when he died we were deported back to Kabul. I didn't want the other migrants to know, I told him. I had more details ready, but that was enough for Zulmay.

"Of course, now it makes sense," he said, sounding relieved. Maybe he felt I was being sincere with him, somehow.

WE WERE CHEERED UP BY some good news: our friend Yousef had escaped. A genial Syrian with a crooked smile, Yousef was trying to rejoin his fiancée, who was a refugee in Sweden. When Omar and I first met him at the port, Yousef had just failed in his fourth escape attempt: first, he'd tried to sneak on the ferry; then, he and a friend were caught in a truck; next, he tried forged asylum papers; and finally, he'd been arrested at the airport with a Bulgarian ID. Lucky for him, the cops usually just let Syrians go.

Then one day I got a message from Yousef: He was in Athens. Yousef didn't speak much English, but he sent me a string of emojis depicting how he'd made it there. He'd bought a ferry ticket and gotten fake papers again—but when it was time to board, it started pouring rain and the cops hadn't even bothered to come outside.

Meanwhile, Omar had been searching high and low for a way out.

There were plenty of people in the camp who had money but the smugglers were having trouble getting them off the island. The boys from the alley had a scheme where we'd hide in a van and then take the ferry; when that fell through, there was talk of a smuggler with a private boat. But nothing came of that, either. The Iranians, demoralized from their failure with the Egyptian, hadn't found anything. There was one option we hadn't tried: Hajji, the same smuggler in Istanbul who'd gotten us into this mess. Our third week in the camp, Omar reluctantly decided to give him a try.

"I was waiting for your call," Hajji said. He apologized for the little misunderstanding on the beach and told us he knew an Athens-based smuggler who could get us out with fake documents for 1,200 euros each. To make it worth his while, Hajji wanted us to get at least four clients together.

"We are four," said Zulmay, once Omar explained.

Omar and I exchanged a look. I'd already decided that using forged papers was a legal line I shouldn't cross. Our plan was that I would call a friend of mine who would bring me my own passport, the unburned one I'd left back in Trieste, so that I could use it to leave the island. Once Omar made it to Athens, I'd follow as Matthieu, and then we could go back underground together. It was still a long way to Italy.

"Habib's not coming," Omar said.

Zulmay and Raja looked so disappointed for me that I felt bad.

"I don't have the money ready yet," I said, "but hopefully I'll get it soon, and then I'll follow."

ONE MORNING WE AWOKE TO find, for the first time, workers making improvements to the camp. They were installing brackets atop the fence that separated our alley from the family compound, so that it could hold coils of razor wire.

There were rumors going around Moria that the holes in the perimeter would be sealed and the camp locked down. Supposedly, now that Afghans could be deported thanks to the new agreement, all of their cases would be processed within ten days; a lucky few would be allowed onward to Athens, but the rest sent back to Afghanistan. One of the boys at the port claimed to have seen mysterious buses arrive in the middle of the night, laden with refugees who were put on the ferry that went to Kavala, in northern Greece.

With winter approaching, people were getting desperate to escape the camp. Some paid smugglers to take them back to Turkey; there were speedboats that after dropping their clients off, picked up others on the island. One day several youthful UNHCR staffers wandered into our alley with clipboards and said they were conducting a census. It seemed that nobody knew exactly how many people were living in Moria. A group of us gathered around an Iranian Danish woman who spoke Persian.

"Why are you doing this?" Omar asked. "Are people being sent to Athens?"

"Well, winter is coming," she said, "so we want to give you nice things."

"Like what?"

"Maybe some warmer tents and blankets, and stuff like that."

After she left, Omar turned to the others and laughed: "We used to make promises like that when I was an interpreter with NATO."

The UN and the government did launch a *winterization* program for Greece's camps. *There are no refugees or migrants living in the cold anymore*, minister Ioannis Mouzalas would proclaim the following January. That same month, an unusual snowfall came to Lesbos, blanketing the little tents. Three men would die, asphyxiated by fumes from the trash they'd burned to stay warm.

No one seemed responsible for Moria as a whole: the police, the UN, the aid agencies, the volunteers, and Frontex all belonged to separate hierarchies. *If one investigated this chain it would possibly lead to thousands of other bosses. All of them repeating the one thing: "The Boss has given orders,"* wrote Boochani of the assemblage at his Australian camp. If no one was answerable, he concluded, *with all his feelings of hopelessness, all a prisoner can do is simply smash his fist against a container wall.* A violence that turned inward, *into a wonderous craving for blood-spill, a wondrous yearning for self-harm.*

Doctors Without Borders would announce a *mental health emergency* on the islands, saying that six to seven migrants were coming to its clinic on Lesbos each week for crises like attempted suicide, self-harm, and psychosis. Of course, people landed here already traumatized from war and the smugglers, but it seemed to me there was also something political in such acts, like the pleas for visibility you heard in news stories, repeated since the border closed: *Why don't they just kill us?* one woman asked. *Just kill me!* a father shouted at the police. *Just shoot me!*

HAJJI CALLED WHILE WE WERE in town and said his guy had arrived from Athens and would meet us in an hour. But there was a hitch: Omar and Zulmay were being sent that evening, but Raja would have to wait for a couple of days so that he could travel with an Iranian family, posing as their child. Zulmay balked but Hajji said it was too risky for the two of them to go together, since someone might ask about the boy's parents.

Zulmay seemed stricken. "I can't leave Raja alone here," he said. He puzzled for a moment, then looked at me. "Habib, I can trust you to look after him, right?"

Oh, shit, I thought, but said: "Of course."

How could I refuse when I was supposedly stuck there without money for a smuggler? And anyway, Moria was no place for a boy to be alone. Zulmay called Raja's mother, who was back in Afghanistan, and explained the situation: Habib, a good Muslim and fellow Afghan, would take care of her child for the two days they'd be separated. Zulmay listened for a minute and then handed me the phone. "She wants to talk to you."

Raja's mother's voice, gravelly from the poor connection, came over speakerphone.

"Hello? Mr. Habib, is that you?"

We exchanged greetings.

"Are you sure you can look after my dear Raja? I'm so worried about him."

"Why, yes I can, *khanom*."

"God bless you, you're like a son to me. Promise me you'll take care of him."

"I promise."

I gave the phone back to Zulmay.

"So it's settled," he said. "Raja will stay with Habib until he goes with the Iranian family to Athens. You're not going to cause trouble for Habib, are you, Raja?"

Raja shook his head.

AFTER A COUPLE OF HOURS, the smuggler from Athens called, and I went with Omar to meet him in a café nearby, posing as a potential client. The smuggler was sitting on the patio in a leather jacket, trim with a shaved head. He was originally from Herat, he said, but had lived in Greece for a decade, and had legal residency now. We were joined by another client of Hajji's, a gaunt young Afghan with bangs that fell past his brow. He'd been stuck with his family in Lesbos for five months and was paying Hajji 4,500 euros to send him all the way to Germany on his own, where he'd send for his wife and kids later.

The smuggler ordered a round of iced cappuccinos, flirting with the waitress in Greek. He passed around a pack of Camels. "I'll give you your plane tickets and documents in a bit," he said.

"We're flying?" Omar said. "What about the ferry?"

He shook his head: "Too many police. The airport is better than the ferry."

The waitress brought our drinks. After she left, the smuggler explained the layout of the airport, and how security would check their bags and documents. When Omar and the others got to Athens, they had to call him and he'd send someone to pick them up at the airport and bring them to the safe house. Omar and a fourth client, an Iranian who wasn't present, would fly out at seven o'clock. Zulmay and the gaunt young father would go on the second flight, at ten.

The smuggler paid for our drinks from a wad of fifties and told us to go wait in the park. Half an hour later, he came back and handed

Omar an envelope with the boarding passes and documents. Hajji had previously asked Omar and Zulmay to send him passport-style photos of themselves against a white background, and the fakes seemed high-quality. There was an Austrian passport for the Afghan going to Germany; a Hungarian ID card for the Iranian; a Spanish one for Zulmay; and with Omar's mug on it, a card from Lithuania.

"Lithuania?" Omar looked at me.

One of the blonder places in Europe, I thought, but held my tongue.

They had the rest of the day to prepare. We'd already gone to one of the Chinese shops that sold cheap clothes, which you could spot by the paper lanterns hung outside. Wandering around the racks, Omar and Zulmay had debated which outfits would look the least migrant-y. A stocky, middle-aged man with a crew cut followed us—a plainclothes cop. Was he just looking for shoplifters, or did he suspect we were planning to escape? We left for another store. There Omar decided on a tight black T-shirt and a pair of black jeans and splurged on a nice roller bag. Zulmay bought a hoodie and decided to stick with his own backpack. To complete their looks, they gelled their hair and shaved.

They changed into their outfits in the park. Omar spent the afternoon memorizing the details on his ID card.

"So, Vygaudas," I asked him in English, "where are you from?"

"I am from Wilnius."

"It's Vilnius."

"Wilnius."

"Vilnius."

"Wilnius."

"Vilnius!"

"I am from . . . Wwwilnius."

"Yeah, that's good. You'll be fine."

We were both nervous. In a few hours, he might be in Athens or he might be in jail.

When it was time, I walked him down to the taxi stand. We passed a rack of used paperbacks in a shop and I bought an English translation of Nikos Kazantzakis's *Report to Greco*.

"Here," I said, handing it to him.

We stopped before the stand, and embraced: "I'll see you in Athens, OK?"

He grinned and set off in his tight black outfit, tugging the roller bag behind him. I walked back to the park and rejoined Zulmay and Raja. We sat on the bench and waited. After half an hour, I figured he must have gone through security. My phone buzzed. It was a message from Omar: *They caught us.*

IT WAS ONLY A TEN-MINUTE drive to the airport along the coastal road, whose lights we'd seen that night setting out in our raft from Turkey. At the entrance, Omar paid the driver and wheeled his bag through the automatic doors. The airport on Lesbos was tiny; during the summer, it was busy with charter flights from around Europe, but it was the off-season now. Omar spotted the Iranian kid ahead of him, and they made furtive eye contact. It was still early, so Omar went outside and had a smoke. Then he went through security; a woman checked his boarding pass and ID and handed them back to him. He walked into the boarding lounge and sat down, thinking he'd made it. But after a minute, he saw the Iranian enter with two uniformed cops. They were heading toward him, and that's when he typed out his message to me: *They caught us.*

But they walked right past with the kid, and into an office. Omar sat there, the paperback unopened in his lap. The minutes ticked by. The plane was boarding. He got up, followed the other passengers out to the bus which drove to the plane on the tarmac. He found his seat. Then he saw the Iranian kid board, with one of the cops behind him, and his heart sank again. But the Iranian took a seat near the front of the plane, the cop got off, and soon they were taxiing to the runway. The kid turned around, spotted him, and flashed a smile and thumbs-up. Confused, Omar pretended not to notice. As they hurtled down the tarmac, he typed out a second message: *I'm on the plane we made it.*

I SQUINTED AT OMAR'S SECOND message. Was it really him or were the police playing some kind of trick? An hour later, he called from the airport in Athens, saying he was about to get on the train.

"He's in Athens!" I shouted, jumping up from the park bench.

Zulmay beamed. "So I have a chance," he said.

Soon it was time for Zulmay to leave for his own flight, leaving me and Raja on the park bench. Now I was responsible for the kid. It was an awkward situation, but I hoped he and Zulmay would soon be reunited in Athens, so that I could follow Omar. My courier friend was already on a ferry to the island with my passport.

"Don't worry, Raja, if you're stuck with me, I'll just ransom you back to your family and then we'll both go to Athens," I said. He smirked.

A young Greek girl rode past on her bike, no hands, with a sideways glance at Raja.

"Can you do a wheelie?" he asked me. "I can."

My phone vibrated. It was Zulmay: *They caught me, by God.*

I sat for a moment, stunned. But I told myself that maybe the same thing would happen as did with Omar, and that Zulmay would somehow make it in the end. In any case, Raja and I had to catch the last bus back to the camp.

We stopped at a gyro shop on the way and I bought a chicken pita for Raja, and a large beer for myself. I wasn't hungry. I watched him pick the fries out, wondering what I'd gotten myself into. When the boy was finished with his pita, I explained what his cousin had written.

"Don't worry, Raja, I'm sure he'll be fine. And I won't leave you until he comes back."

He nodded, and looked across the harbor, toward the dark space where the ferry tied up during the day.

When we arrived at the camp, I saw a bunch of cops at the entrance, so we went up the hill and slipped in through the holes. I was worried that if the authorities knew that Raja's guardian had been arrested, they might try to lock him up in the children's area, for his own protection. But as long as I was around, no one was going to touch this kid. Moria seemed more on edge than usual; knots of men stood in the dark spaces between floodlights. Back at the *kucha*, we heard about the day's demonstrations and clashes with the police, after it was announced that asylum interviews had been postponed by two more months. Some Eritreans had been arrested.

I put Raja to bed in his tent and then crawled into mine; it felt strangely spacious, on my own. I wondered how Omar was doing in Athens, but his phone was off. And what had become of Zulmay? A coughing fit seized me; my cold seemed to be getting worse.

WE ROSE EARLY AND TOOK the bus into town. I brought my backpack with me, in case we didn't come back that night. There was no news and I assumed Zulmay was in jail. He was probably facing a six-month sentence. I was anxious to get to Athens, but I couldn't leave Raja on his own. I was basically on the lam with a child, but I'd promised his mother that I'd look after him. Anyhow, it appeared that Hajji still planned to send Raja with the Iranian family to Athens.

The boy was taking things calmly. We sat in the park while he fielded calls from his mother and Hajji. "She wants to see you," Raja asked, and then swiveled the phone's camera toward me. I waved and tried to smile reassuringly.

Hajji kept leaving rambling voicemails for Raja on Viber: "Don't worry, don't be scared, I'll get you out. It was Zulmay's mistake that got him caught, there goes all my expenses, but never mind, you'll go to Athens and I guess you'll be on your own there." Hajji giggled. "But don't be scared. From there we can send you to Germany, if that's where you want to go."

Raja held the phone up and left his reply: "No problem, Hajji, I understand, whatever you say, you know better than I do. Just please call my mother and explain it to her."

To kill time, we went for a walk along the docks, peering at the old tankers tied up by the container port. Farther on was a traveling fair with amusement rides, and beside it a Roma encampment, with tents that looked like the ones from Moria, and a couple of lived-in sedans. A little girl carried an infant past us, as an elderly woman stirred a pot that smelled of beans and onions.

We sat down on a bench nearby. One of the minor hardships of this trip was having to forswear English reading material, since it wasn't in character. But with Omar and Zulmay gone, I had broken down and bought a novel in town, *The Days of Abandonment* by Elena Ferrante.

I was about halfway through when I looked up and saw Raja kicking at the bench and staring into space. The poor kid was bored. But what could we do together? An idea came to me.

Earlier, I'd noticed a sign for an Internet café near Sappho Square. Inside we found a purple-lit cave with rows of computers. Behind the counter, a girl in an anime T-shirt was chatting with a trio of dudes. She eyed us skeptically as we entered.

"Do you have any computer games?" I asked in English.

She frowned and waved her cigarette: "We have more than five hundred."

It had been a while since my gaming days. No CD-ROMs were involved anymore, it seemed. Everything was on the cloud. I rented us two computers next to each other and we made our selections. Raja ran amok through Los Angeles in a stolen Lamborghini with a loudmouth Lindsay Lohan lookalike in the passenger seat, while I shot at Talibs with a SCAR chambered in 7.62 mm NATO.

Hours passed in this manner. The smoky air was burning my inflamed lungs. I was sweating, although I had the chills. It was getting late, and Hajji had stopped responding to Raja's messages. I didn't feel safe bringing him back to Moria. I checked my phone: my friend had arrived on the island with my passport. I didn't want to surprise her with Raja, but I couldn't exactly take the boy to a hotel on my own, either. Yet I needed to rest; my head felt like it was about to fall off. I was wondering what to do when Zulmay texted: He was out of jail.

THE SMUGGLERS HAD THIS HABIT of sending their clients in groups. I guess it was because some refugees had trouble navigating the airport, so it was better to make sure at least one person spoke English and knew what to do. But the result was that if one was busted, the others usually were too.

Zulmay had made it through security and figured he was home free when he heard a shout behind him. A cop came up and hand-cuffed him right there in the boarding lounge. He took him to a room and left, locking the door behind him. Sitting there was the other Afghan, the young father with the bangs.

"Did you sell me out?" Zulmay demanded.

"I swear to God I didn't say anything."

A cop in plainclothes came in and asked Zulmay in English where he was from.

"Spain," Zulmay said.

The guy laughed. "I've caught a thousand people like you." He started speaking to Zulmay in Spanish.

They were put in a jail until the following afternoon, then taken to court. The other Afghan, who'd worked for foreign NGOs in Kabul and spoke good English, asked the cop escorting them what would happen.

"It depends on the judge but you'll probably get six months."

"What?" The Afghan started pleading. "You promised you'd help me!"

The cop shrugged. "Tell that to the judge."

So he was a snitch, thought Zulmay, seething.

They were led into the courtroom still cuffed. The judge had an interpreter, a woman whose English Zulmay found hard to follow. Ahead of them, there were a young African man and a Greek woman who'd been caught at the ferry the night before. After hearing the details of the case, the judge sentenced the man to six months, while the woman got a year for smuggling. They were led out in tears.

The judge told Zulmay to explain himself. He answered that he'd just wanted to get his cousin out of the awful conditions in the camp, so they could join their family in Germany. He asked the judge not to imprison him for Raja's sake. The judge chastised him: How could he leave a child behind like that?

Zulmay hung his head, as he had before the Knight of Rhodes.

Through the interpreter, the judge announced that he was taking into consideration that it was their first offense, and that they'd come from a war-torn country. He was giving them each a six-month sentence and a 1,500-euro fine, and if they were caught again they'd get a year in prison.

Zulmay felt the room spin. What would happen to Raja?

The other Afghan started wailing: "You can't do this to me! Please, I'm a civil society activist."

At this outburst, the judge looked confused and conferred with the interpreter. Embarrassed, she clarified that they had been given suspended sentences. They wouldn't go to jail unless they were caught again.

That night, the cops took them back to the camp and let them go. Zulmay had been looking for us in the *kucha*; over the phone, I told him to come out and wait at the bus stop by the gate, because I was sending Raja to him—it was a short, direct ride.

I walked with the boy down to Sappho Square, and we waited for the bus. When it came, Raja started climbing the steps, and then turned back to where I was standing.

"You're not coming?" he said, confused.

"Not right now," I said. "Give my regards to Zulmay."

"I'll see you soon," he said, and smiled drowsily. He gave me a hug, and got on. I shivered; the streetlights seemed to flicker. I waited there until the bus drove out of sight, taking the boy back to Moria.

Part IV

The City

17

Before you came things were just what they were, wrote the poet Faiz Ahmed Faiz: *the road precisely a road, the horizon fixed, the limit of what could be seen, a glass of wine was no more than a glass of wine.*

Omar looked at the water far below, its white scars fading in the twilight. This was his first time in an airplane over the sea. Laila had never flown at all, but one day, he promised himself, he would put her on a plane to Europe and take her away from the war. Omar's flight crossed the Aegean in less than an hour. His stomach clenched, feeling the plane descend. He'd escaped the island, but he was afraid there would be police checks at the airport in Athens. If they found out he'd been in Moria, they would send him back. He felt in his pocket for the plastic card. He was a Lithuanian named Vygaudas.

Omar followed the other passengers out of the airport, his gaze straight ahead, through the sliding doors to where he inhaled the cool air and saw that night had fallen. He called me to say he'd arrived in Athens, and then started looking for the train. He needed to get away from the airport. His phone rang. It was the smuggler. Omar was supposed to go to the man's safe house in Athens, because they wanted the Lithuanian ID back. But why should he return it? The smuggler had claimed it belonged to another person but that made no sense, since the card had Omar's picture on it. And he'd paid for it. They probably just wanted to make him buy documents again in

order to leave Greece. But he needed the card here in Athens. He ignored the call.

The signs led him up an escalator and over a pedestrian bridge. He
bought a ticket from the machine, and when the train came, found a
seat among the travelers and luggage. No one paid him any attention.
The smuggler kept calling until the train entered the city center and
went underground. He transferred lines, and when he heard his stop,
Victoria, he got off.

Victoria Square was the hub for Afghan migrants in Athens. Omar
climbed the stairs of the station and saw a rectangular park, two by
four blocks, with an enormous bronze sculpture in the center atop a
graffitied plinth, a centaur abducting a struggling, nude woman, now
streaked with pigeon shit. Groups of men sat on the wooden benches
around it, conversing in Persian. Omar asked around and quickly
found what he was looking for: a *khobgah*.

The apartment blocks around Victoria were filled with migrant
flophouses where five euros a night got you a place on the couch or
floor. No one asked if you had documents. The one Omar found had a
pile of flip-flops and sneakers at the door, the stained carpet bare apart
from a hot plate and a snarl of phones by a power strip, with a fuzzy
TV in the corner, the bathroom layered with human sediment, the air
thick with cigarette smoke and languid conversations about smugglers
and the old country.

The occupants were Afghan: two families in the bedrooms in back,
and eight men in the living room, already lined up under fleece blankets in the dark by the time he got there, some faces lit by phone light.
As he lay down, his Samsung rang. This time, he answered.

"What happened? Where are you?" It was the smuggler he met on
Lesbos, the one with the shaved head.

"I went to a friend's house in Athens. Sorry."

"We need the card back."

"I got scared on the plane that the police were going to catch me,"
Omar answered, "so I threw it in the bathroom garbage."

"Is that so? I don't believe you. Why don't you come over here?"

"Maybe tomorrow," Omar said, hanging up. Then he saw a message from Hajji: *You'd better give the card back, or it won't end well for you.*

Fuck you, he thought, and turned off his phone.

WHEN HE WOKE, OMAR LEFT the musty apartment and went down to the square, seeing the neighborhood in daylight, the grimy apartment buildings and cracked sidewalks reeking of urine. There was an Afghan restaurant on the corner, and several Chinese shops with plastic goods in the window. So this was the famous Victoria, Omar thought; it was known to Afghan migrants as a key that would, like Zeytinburnu in Istanbul, decrypt a foreign city. He called me and learned that Zulmay had been arrested, and that I was stuck with Raja in Lesbos. He would have to wait for me, I told him. Omar sat down on one of the park benches, next to an elderly Greek. So Zulmay was the unlucky one, he thought. The odds had been fifty-fifty.

He called Maryam to let her know that he had made it safely to Athens, then explored the square, smoked, sat on a few different benches, had a sandwich, wandered and smoked some more. By late afternoon, he was bored enough to ask another Afghan if there was a park or something nearby, a place where he could relax and enjoy himself.

"You mean Alexander Park?" the man said. He pointed east. "Just keep walking that way." Omar crossed a busy intersection and saw, at the edge of a sprawling green space, another giant equestrian statue, this one of King Constantine I, whom the Afghans assumed was Alexander the Great. Maybe they'll have amusement rides in the park, like in Iran, thought Omar. On the other side of the statue, he saw tents among the trees. Had some tourists camped here? As he drew closer he saw that the people there, unkempt and shabbily dressed, were smoking powder off tinfoil. There was a tall man standing on the path, and Omar watched him make a sale.

"What do you sell here?" he asked in English.

"Heroin and *shisha*," the man answered, in a thick accent. *Shisha* was methamphetamine.

"Who buys it from you?"

"Everyone," the dealer said, then sized up Omar's broad frame. "What are you, police?"

"Why would I speak English if I was police?"

Baffled, the man stared at him, and then snarled: "If you want to buy, then buy, otherwise go, don't ask questions."

Omar continued down the path, noting the heaps of garbage amid the scruffy vegetation. He approached a group of young men speaking Pashto, and exchanged greetings. The men explained that they'd been stuck in Greece since the Macedonian border closed eight months ago.

"But what happened?" Omar asked, seeing how gaunt their faces were. "How did you become addicted?"

They looked at each other and smirked.

"Our money is gone, we have nothing to do."

"We have mental problems, brother."

"So how can you pay for drugs?" he retorted. They looked embarrassed.

"God provides," one answered.

"We find it somehow. Some boys steal."

They pointed to a younger Afghan boy standing nearby.

"He's a *kuni*. He sells his ass."

"You're new here—you'll see."

Omar said farewell, and walked deeper in the park. It was shameful to see his countrymen like that. How could this be happening in Europe? He knew that Greece was having an economic crisis, but he hadn't imagined that Athens would be this run down. Istanbul seemed so modern in comparison. No wonder migrants didn't want to be stuck here.

He sat down on a cracked bench. The bushes were thicker in this part of the park. There were a few other men sitting nearby, singly or in pairs, some eyeing him. At least it was calm here, Omar thought, puffing on a Winston.

A young transwoman—he'd met others before in Pakistan, where third-gendered *hijra* were part of traditional culture—teetered up on heels and asked, in English, for a cigarette.

"Do you want sex?" she said, after he lit it for her.

"No thanks," he answered glumly.

Night was falling, and though there weren't many lamps, this area was getting busier. Some of the pairs of men went off into the bushes, and others would take their place on the benches. A chubby Greek sat down next to him.

"Hello. Where are you from?" he asked.

"Lithuania," Omar said.

The man seemed puzzled by this. "So, what do you like? Do you like sex?"

"Of course I like sex."

"I can suck you off if you want."

Omar laughed. "No, I meant with a woman."

The man hustled off. It was quite dark now, and Omar decided to leave. He headed back to Victoria Square, marveling at some of the couples he passed on the streets: young women with decrepit men. He saw a girl sitting on the sidewalk, with her legs akimbo.

"Hello," he said, but she just stared ahead vacantly.

It was early, but he felt tired. He went back to the flophouse.

WHEN THE BUS CARRYING RAJA back to Moria had rounded the corner, I called the friend who'd brought my passport with her to Lesbos. It was too late to leave for Athens that night, so she found us a place in town with someone she knew. Whatever bug I'd caught in the camp was bad: I went to bed wracked by chills and coughing fits and woke up in the middle of the night, skull like an inflating balloon. The next couple of days passed in a delirium I can't really remember apart from painful visions of impossible geometric figures, and my friend's hand on my brow, her voice announcing my temperature, saying that after a couple more degrees we'd have to go to the hospital, but in the end the fever broke.

As soon as I could walk, I was anxious to get to Athens. My friend was worried for me but I thought I might get her in trouble if we traveled together so I went to the airport alone, pallid and clean-shaven, avoiding eye contact with a cop I recognized from the camp. Passport and ticket in hand, I boarded my flight.

In Victoria Park, Omar and I embraced, savoring for a moment our escape from the island. But now we had to decide what to do next. I looked around the square, at the loitering touts for smugglers. The streets of Athens weren't particularly safe. If the cops stopped Omar and figured out he'd escaped from Moria, he might be sent back or jailed on the mainland. And I hadn't warned him yet about the fascist vigilantes that roamed the city. Ten minutes away from us was Saint Panteleimon Church, where the Golden Dawn, who marched in black shirts under a swastika-like banner, had led a riot against immigrants, after some Afghans killed a local man during a robbery. We were better off moving onward into the underground, but first we had to figure out how. It was a long way to Italy through the Balkans, and I was exhausted from my illness. We needed to rest and think things over.

"I still have the card," Omar said. "Maybe we can get a hotel with it."

"What card?"

I listened in dismay as he explained how he'd absconded from the smugglers. The *khobgah* were sketchy as it was, not to mention unsanitary; now we had to worry about Hajji and his thugs. A hotel was no good because of the police, but I had another idea.

I'd come to Athens for the first time almost two years earlier, for the national elections in January 2015. After Greeks had voted in protest of the harsh austerity imposed after the debt crisis, Syriza, a radical socialist party, won a surprise victory. It was a euphoric moment for the Greek left, a time of great hope, and during the celebrations I had met some local No Borders activists, who were just the kind of people who'd be sympathetic to our current predicament. So I rang up my friend Nasim, an Afghan who had settled as a refugee in Greece, and told him that I'd just arrived in town, and needed to see him about something important. He said to come to Exarchia.

The neighborhood was an anarchist stronghold only fifteen minutes away on foot, but to get there we had to cross the front line. At the Archeological Museum, a group of burly riot police stood guard beside their bus as we walked past into the autonomous zone. Elefthera Exarchia, some residents called it: Free Exarchia. Long associated with radical politics, the neighborhood became a no-go zone for the police in 2008, when they gunned a fifteen-year-old boy, Alexis Grigoropoulos. The city exploded in recurring protests that intensified the following year, when the global financial crisis pitched Greece into a depression worse than that of the 1930s.

The buildings were covered in graffiti and posters, a tangle that grew denser as we approached the central square. The walls were a message board for the neighorhood, announcing punk festivals and fundraisers for jailed activists; some bore portraits of antifascists killed fighting with the Kurds against ISIS, others had spray-painted slogans in Greek and English like ALL COPS ARE BASTARDS and NO ONE IS ILLEGAL. At last we arrived at Exarchia Square, in truth a triangle with scraggly trees; the Greek word, *plateia*, denoted in ancient times a wide place of assembly—square like Tahrir or Maidan, Tiananmen or Zuccotti. A banner at one corner announced in English: AIRBNB SUPPORTERS GO HOME, HERE WE HAVE CLASS WAR.

We were safer in Exarchia. The Golden Dawn's vigilantes knew better than to come here, and even the cops could only enter in force, if they were prepared for a brawl with Molotov-wielding youths who defended streets like a medieval casbah's, narrow enough in spots to jump from roof to roof. The anarchists lived here; the neighborhood was full of occupied venues, collective cafés, and squats, even a self-managed park, Navarinou, where the people had torn up a parking lot and planted trees: *Sous les pavés, la plage.*

We were meeting Nasim at Steki Metanaston, a social center for leftist groups including his own, Diktyo, a network of immigration activists around Greece that had been working for decades. Steki occupied a three-story townhouse just off the square, up a steep, pedestrianized alley. Omar and I climbed the stoop to the main floor, where a few people sat smoking at battered tables by a bar, whose cabinet displayed a mismatched collection of mugs, jiggers, and glass carafes for raki, the ubiquitous Greek spirit. The cupboards and walls were covered with posters and stickers, some from foreign ska bands and syndicalists that had passed through and others from the *kinema allilengyis*, the Greek solidarity movement, an efflorescence of volunteer clinics, food banks, free schools, and other grassroots responses to the economic crisis. When the little boats came ashore the previous year, many solidarity groups started helping refugees in concert with organizations like Diktyo, merging into something the activists simply called the Movement.

In the back room, beneath a pastel mural painted by a visiting Zapatista group, a meeting was underway, and among the heads I spotted Nasim's wiry black pompadour. When he saw me, he ducked out and grabbed three bottles of Alfa from the bar. We sat down and clinked our beers, grinning at each other; to be honest, I found some of the Athenian activists a bit intimidating, with their acid contempt for the bourgeoisie, but Nasim and I got along easily. He was from Jaghori, a Hazara district in central Afghanistan that had produced an extraordinarily wide and talented diaspora, but he'd not seen his homeland since he was a child. He had grown up as a refugee in Iran, and fled to Turkey as a teenager. About ten years earlier, he'd paddled to Lesbos in a rubber raft with three friends. By now he was thoroughly Hellenicized and radicalized, a chain-smoking raki drinker who spoke fluent Greek and English and was one of Diktyo's most visible spokespersons. That decade of street activism and hard living had taken its toll. He looked older than his thirty-odd years, and wore his usual grim expression as he listened to our story, but when I told him how we'd taken the boat from Turkey and ended up in Moria, his boyish smile shone through again.

"You're crazy," he said, chuckling. He had just come back from the islands himself, on a gig translating for a human rights group that was documenting the appalling conditions in the camps. "Moria is the worst," he agreed.

I told Nasim that we needed shelter while we figured out our next move. Could they take us in? The previous spring, Diktyo and some other groups had occupied an empty hotel near Victoria, cutting the locks and moving in refugees from the camps and streets. There were around four hundred people living in the squat now, a mix of activists and migrants.

Nasim frowned at his Marlboro and rubbed his brow, his eyes

bloodshot with fatigue. "The problem is that there is a long waiting list," he said. Families were given priority. However, they also had rooms for foreign volunteers visiting Athens. He could get us in that way, but we'd have to work, help out in the kitchen, that sort of thing.

"Come tomorrow afternoon. I'll check you in," Nasim said. He drained his Alfa and went back to the meeting.

THE HOTEL WAS A FIVE-MINUTE walk from Victoria Square, a stolid, eight-story concrete building with a patchwork of laundry hanging from the balconies. A sign above the sidewalk read:

H
O
T
E
L

C
I
T
Y

P
L
A
Z
A

The honeymoon between the leftist Syriza government and the Movement hadn't lasted long. Within seven months of the election, Greece's creditors, led by Germany, broke Syriza's resistance to austerity with the threat of being kicked out of the euro. Then came Europe's refugee crisis; Syriza, which had promised to end the mass incarceration of migrants, was forced to turn Lesbos and the other islands into prisons. Diktyo and the other activists who'd supported Syriza were horrified. That past winter, when the borders in the Balkans were closed, some fifty thousand refugees had been trapped in mainland Greece, many living in the streets or makeshift camps like the one at the decommissioned Ellinikon Airport, where families slept under analog boards still listing flights to London and Frankfurt. Athens was full of buildings abandoned during the economic crisis; in the face of the humanitarian crisis taking place on the streets of their city, the activists had defied the government and helped refugees occupy them. By now there were around two thousand migrants living in a few dozen squats in and around Exarchia.

City Plaza was the biggest and the only one I knew of where the activists and refugees lived together. So far, the government had left them alone, despite the protests of the owner, an actor whose father established the now bankrupt hotel in 1974. The Greek media were another story. If migrants trespass against the nation, squatters violate something even more sacred, private property, so the migrant squatter was a reliable bogeyman for Athens's conservative press, much of it owned by the country's oligarchs. To read the Greek papers, you'd think there was something sinister happening inside the squatted hotel that Omar and I now approached.

The entrance was on a side street. We told the activists manning the security desk we were here to see Nasim, and went up into the lobby decorated with abstract prints and a marble fountain. Behind the counter, Nasim and several black-clad Greeks were smoking hand-rolled cigarettes and parleying with a Syrian couple who wanted more diapers. Spotting us, he grabbed a key from the rack and we followed him upstairs past the dining room, which smelled of onions and steamed rice, and then up the main stairwell. Nasim explained that the squat was a democratic collective, run by consensus, where everybody had to work but there were no bosses, which was hard for some people to understand. The important rules to remember were no violence and no alcohol—the latter partly out of respect for Muslim residents, and also just to avoid trouble.

We kept climbing. There were two elevators, but Nasim said they'd been deactivated because the children kept playing with them. There were nearly a hundred refugee kids living in the squat, and many had grown wild in the year they'd spent following their parents into smugglers' dens and across the water. The kids, who liked to play in the stairwell, rushed past us in groups, shouting in Dari and Arabic, their squeals echoing off the tiled surfaces.

Our legs were burning as we arrived at the sixth, penultimate floor, where most of the volunteers lived. Nasim unlocked our room, handed us the key, and said he'd fill us in later on how the shifts worked.

Omar and I entered and set our backpacks down. The room was painted cream, trimmed with blond wood, and had its own bathroom. A nightstand separated two single beds. A sliding glass door opened to a tiny balcony, and we went out to look down at the streets we'd been roaming a moment earlier. We grinned at each other. The previous occupants might have left it messy, but these were the nicest digs we'd stayed in since leaving Kabul.

"I'll go ask for a broom," Omar said.

AFTER A COUPLE OF NIGHTS in a hotel bed, I was feeling better. Hajji and his henchmen had stopped calling, and Omar was confident enough to go down to Victoria Square and ask around about the road ahead. I stayed back to lounge in the room, enjoying the solitude. The journey through the Balkans was sure to be brutal; it would be nice to stay awhile in the squat, if only winter wasn't so close.

There was a sharp knock on the door. I jumped to my feet and opened it to an Afghan man in his midforties, with the neat black beard of a sea captain, dressed in a denim jacket and a pair of camouflage pants.

"Fatiha," he announced gruffly, and moved on to the next room.

The Sura al-Fatiha, the first chapter of the Quran, was an Afghan shorthand for the prayer service held after someone's death. The man had assumed I was a refugee, too. Omar and I were an ambiguous pair at Plaza. Only Nasim knew the secret of our escape from Lesbos, and whenever people asked I introduced myself as a Canadian journalist named Matthieu, since some of the activists knew me already. But most of the other refugees, hearing Omar and me speak together in Dari, assumed we were fresh off the boat. That was just as well, because outside the squat I still had to be Habib, in preparation for our return to the underground.

I put my shoes on and descended the six flights of stairs to the mezzanine, the main communal space which held the dining room, bar, and kitchen. The walls were covered in children's art, hand-drawn schedules, and a grid of portraits taken by visiting photographers, a bricolage reminiscent of a hip elementary school. I followed the sound of Quranic recitation to the conference room, where the fifty-odd chairs lining the walls were occupied by men and boys—there was a separate ceremony for women—who greeted me, "*As-salamu alaykum*," as I joined. We sat and listened to the classical Arabic keening from a boom box: *All praise be to God, Lord of all the worlds.* I looked from face to face: Arabs, Afghans, Kurds, and Pakistanis, unknown to one another six months ago and now gathered to mourn a man from Herat named Waleed, who had drowned the day before on the outskirts of Athens.

Hamid, a young Afghan with a skinny frame and bushy eyebrows, came around with a box of dates and cups of steaming, sweetened tea. Later he'd tell me about his friend Waleed, who'd come from Afghanistan with his daughter and wife, hoping to join her family, who were already legal residents of Switzerland. Like the rest of them, Waleed had struggled with despair when the border shut, trapping him in Greece. Two days earlier, he'd gone to the beach alone at night; he was a strong swimmer. The police had found his body in the morning.

Hamid clicked off the stereo, and we went through rounds of prayer, our upturned hands sweeping our faces each time in unison. "Truly, we belong to God and to him we shall return."

Afterward, Waleed's brother-in-law, who'd flown in from Switzerland for the occasion, thanked everyone. There'd been talk by the family of burying Waleed here in Greece, but in the end they found the money to fly him back to his native Herat, where he would lie amid the wildflowers he'd known as a child.

We shuffled out, and I saw Nasim, downcast and pensive. It was hard to understand, he said to me afterward, how someone could survive decades of war, walk through those mountains, cross the sea in a raft, only to lose his life like this in Europe. "There is a saying in Greek," Nasim said. "The one born to drown won't die another way."

AFTER A WEEK OF BLUE skies, dark clouds piled up over the ridges and burst open thunderously, raining *chair legs*, as Athenians said, until the gutters overflowed. Omar and I were glad to have a roof over our heads this time; then, a couple of hours into the downpour, a stream of water burst through the drop ceiling in the mezzanine, the leak somehow bypassing the top six floors to shower the residents who'd lined up for lunch. Omar and I were working in the kitchen and a request came for pots and pans to catch the rain.

Having yet to figure out our next move, Omar and I had been absorbed into the hectic life of the squat, helping where we could with the security and maintenance shifts, translating for the clinic staffed by volunteer doctors, our activities coordinated by the team at the reception desk who fielded daily crises in a babel of tongues and a Mediterranean mix of mannerisms: the plaintive interrogative with upturned fingertips, *Why*; the silent chin-hitch negation, *No*; and the ubiquitous vocative: *Filemou, habibi, my friend!*

The most important task was running the kitchen, for even an army of lovers marches on its stomach. Due to ongoing legal disputes, the bankrupt hotel still had all its equipment. Ringed by cauldrons and an industrial-size oven, three gas burners, and a hand-cranked cooking tub which resembled a medieval siege weapon, one of three rotating chefs—an Arab, a Kurd, and a Greco-Chilean—clenched fistfuls of cumin and paprika and swung an immersion blender the length of a baseball bat, while a dozen of us minions chopped forty-pound bagfuls of onions and potatoes we'd hauled out of the squat's battered van, bushels of tomatoes and lettuce, assembling utilitarian stews of lentils and beans with rice or pita bread, the menu dependent on what had been cheap at the farmers' market, or was donated. Once we received half a ton of cucumbers and raced to consume them before they went moldy, serving them in salads and yogurts and just straight up.

Toddlers wandered the halls gnawing gourds the size of their arms, and for weeks afterward we kept finding cukes secreted behind the furniture. Sometimes the food was undercooked or scorched, but sometimes, my friend, it was sublime. Baked meatballs spiked with rice in garlicky tomato sauce. Fattoush salad with homemade croutons and an emulsion of lemon juice and local olive oil. The Syrian chef whom everyone called Mama made rice in the Levantine-Persian style, cooked with a little oil in the water so that the bottom layer turned golden crisp, and Omar's eyes bulged to see the platter-size disks of *tahdig* she pulled at the end and fed to her helpers.

Mealtime brought a clamoring, hungry throng down to the big dining room. Some residents filled containers and went back up to their rooms while others sat together at the round tables. Two-thirds of the way through we had to collect dishes and run them through the Omniwash to reuse them. That was when people started to sneak off the kitchen shift. Cleanup was left to the cook and whoever stuck around; there'd always be a couple of earnest souls left behind scraping out the cauldrons.

The day after the rainstorm, Nasim came into the bar, as everyone called it, though it only served treacly concoctions of Nescafé, and asked the volunteers if we could go up on the roof and find the leak. Maybe clean up the deck while we were at it, he added. Alongside the activists from groups like Diktyo, who were largely Greek, and the refugees, the volunteers formed the third leg of Plaza's tripod. They were a motley crew of more or less ordinary people, most of them young, visiting from around Europe and farther afield, who helped out in the squat, often in exchange for a free place to stay. The volunteers were drawn here by word of mouth, social media, and news stories, for Plaza had a savvy press team and got admiring coverage from the foreign, if not Greek, media: *Time* magazine had described the squat as *a virtual paradise compared to the grim conditions of the government run camps that house most migrants.*

As per our agreement with Nasim, Omar and I had enlisted among the volunteers, and since we both spoke English, they assumed that we were, like them, legal visitors to Athens. It wasn't unusual for the volunteers to have backgrounds from the same countries as the refugees, like Zied from Tunisia. Others, like Carles the Catalan who'd busked to Brazil and back, were nearly as broke. The line between us was drawn by our documents.

At Nasim's request, a few of us set off upstairs with some Afghan and Syrian teenagers that he'd dragooned as well. At the top, Mar, who was from Valencia, unlocked the door and we walked onto the rooftop patio, an expanse strewn with broken lawn furniture. In the corners, the windswept dirt was piled high enough for weeds to sprout. The menu was still up beside the bar; six years ago, guests had sipped Campari and nibbled toasts here. Mar and the teenagers started sweeping up, unearthing condoms and beer bottles from the building's derelict phase. Accompanied by Henrique, a strapping Portuguese backpacker en route to Sydney, I tried to find the leak. The other half of the roof housed the air-conditioning and ventilation systems, along with two massive, German-made elevator motors. We sent someone downstairs with a walkie-talkie, and hosed water into the elevator and cable shafts, trying to determine if they were the source, but found nothing.

Afterward, the teens drifted off, and the volunteers climbed onto the terra-cotta roof where we passed a spliff around, the smoke curling and catching the light. The hotel was the tallest building in the neighborhood, and we had a panoramic view. To the southeast were the forested slopes of Mount Lycabettus, with the dome of Saint George's chapel atop. Farther south, we could see the white marble columns of the Parthenon. The faraway port was obscured by haze; inland, the horizon was hemmed by bare mountains. The apartment blocks extended in all directions, wrapped in awnings and balconies and sprouting satellite dishes and water tanks.

The sun was hot on our bare limbs as the volunteers talked idly about where we were from and where we were going. Some had come to Athens to take part in City Plaza, while others just stumbled upon it while on vacation. The strange thing about being myself among the volunteers was that it was a cover for the fact that I was here in Greece illegally. I had to be coy sometimes because they were always asking one another questions, biographical mainly but also the big ones, like

why we cared about the refugees and what the solution to the migration crisis was. What ought we to do in a world of strangers drowning?
We lacked the activists' certainties. The City Plaza project called for
open borders, but it wasn't easy to see how that would work. Words
like *democracy* and *justice* sounded right but led you in circles when
you scaled them up to the planet, their commandments making ring
fences around the nation-state. The volunteers were searching for
truth, but at Plaza they found love.

Now we were out of the camp and less concerned with survival,
Omar thought of Laila often. He hadn't spoken to her since he left Kabul, and tormented himself by imagining her surrounded by suitors,
like Penelope. He was impotent against them. To ease his mind, I'd
try to get him to come out with us, on nights when a mixed group of
Greeks, volunteers, and refugees would traipse down to Exarchia. We
walked there past grand, crumbling houses, their courtyards shaggy
with bougainvillea, the wrought iron rusted and the marble chipped
but still *good bones*, as the developers would say. Despite the crisis
which had cut Greeks' income by 40 percent, property values and rents
were going up in central Athens, as foreign investors bought apart-
ments and rented them to tourists on Airbnb. As an added incentive,
Greece had the cheapest golden visa in the EU, residency in return for
purchasing a quarter million euros in real estate, and you often saw
groups of Chinese buyers touring the city in coach buses.

Our crew passed under the gaze of the riot police and into the autonomous zone. In those days, the square was always busy, especially
on weekends, when kids came from around Athens to drink cheap
beer from the kiosks or at one of the techno parties the anarchists threw
to raise money. They were joined by young migrants from the squats
around the neighborhood, some like the one on Notara Street well
supported by locals, but others grim setups in gutted buildings. The
crowd mixed harmoniously, but beneath the surface Free Exarchia was
coming apart under outside pressure and the weight of its own con-
tradictions. For years, the neighborhood had been besieged by drug

gangs, who wanted to sell to the partygoers and were secretly encouraged, according to the anarchists, by the police. The trouble was the gangs now were recruiting among the migrants from the squats. Earlier that year, an Egyptian dealer stabbed some anarchists in front of their café, which had already been shot up once. Afterward, squads of baton-wielding anarchists cleared the square, beating up anyone they suspected of dealing; later, they staged an armed march through the neighborhood. When the same dealer had unwisely returned, he was shot in the head near the square. *We take the responsibility for the execution of mafioso Habibi*, announced the previously unknown Armed Militia Groups. *These scum, who pretend to be Escobar and fearless, are common snitches and associates of the police. . . .*

But none of this violence was evident those mild fall nights, not when the square was filled with music and the hum of voices. The City Plaza crew would drink at the tables outside the tattoo parlor run by the Antifa Club, the group of radicalized ex-hooligans who ran security at Plaza. They had a fight club, so I heard, and some trained with an Afghan kung fu black belt who'd broken a fascist's arm in Victoria. Migrantifa, Nasim termed it. We ended the night next door at Steki, the leftist clubhouse where, no matter the night, the DJ was obliged to play certain tunes; we toasted each other to "Bella Ciao," the partisan's anthem, and then to Manu: *Mi vida va prohibida, dice la autoridad.*

People from the squat kept straggling in to acclaim, Chef Shero and his Kurdish crew, Australian Ned who could dance in a handstand, even Omar, who nursed an Alfa and smiled back at the women but thought of Laila. Romance was in the air at Plaza, with so many young people living in proximity. They were only human, and skin was just skin, the last border between two hearts. Quite a few of the female volunteers—although not vice versa—got together with male refugees, which might have been a scandal at an NGO but not in a squat where everyone was meant to be equal; and, on some nights at Steki, it felt like we were.

"Look, do you see that little girl in the photo?" said Hamid, who'd served sweets at the mourning ceremony. We were standing in the mezzanine under the black-and-white portraits taken by visitors. He pointed to a child with curls and a radiant smile. "Her whole family made it to Sweden on the first try. Six people."

They had paid more than ten thousand euros for fake documents and flew to Stockholm, hitting jackpot in the game that half the squat was playing. Refugees got makeovers to look more European and disappeared, gone with the smugglers. The next time you saw them was on Facebook in Germany. Or else they came back to the squat a few days later, busted and looking sheepish in their new haircut. Or they went to jail.

In liberal democracies, the border has a unique power to transmute ordinary needs into criminal desires. The Greeks were in this for the long run and made it their business not to know about such smuggling intrigues, but some volunteers got drawn in. It wasn't because of some new truth they'd discovered; rather, the refugee crisis had become the crisis of friends and lovers. Their problem was a line, and the solution was to cross it. And so some volunteers gave their own documents to refugees who resembled them; some mixed couples left Greece together, posing as tourists. These were among the *crimes of solidarity* committed that year across Europe. In Italy and Greece, volunteer boat rescuers were being prosecuted as smugglers. A seventy-year-old Danish woman was fined for giving a ride to migrants walking the highway. In France, a farmer was convicted for helping hundreds crossing the mountainous border with Italy, in what the papers called a *French Underground Railroad*; his sentence was overturned on appeal, under the constitutional principle of *fraternité*. That fall, an activist from Barcelona was arrested at the airport in Athens for trying to bring a Kurdish teenager on the plane with her. She'd given the boy her own son's documents.

Late one night, the widow of Waleed, the drowned man, found me in the mezzanine. She had her little daughter in tow, and asked if there was a child's backpack among the donated items. There were regular hours for clothing distribution, but she needed it that same night. I didn't ask why. We opened up the storeroom and sifted around. That was the last time I saw her but I heard later that she and her daughter made it to Switzerland.

18

Each time I stepped outside of Plaza, I had to remind myself that I was Habib. In Victoria Square, Omar and I kept running into refugees we knew from Lesbos, who'd either escaped or been granted an asylum interview in the capital, such as Yousef, the genial Syrian who'd snuck onto the ferry on his fifth try. The Arabs usually stuck to Syntagma Square downtown, but Yousef would drop by Victoria to chat with the Afghan and Pakistani smugglers who ran people over the border into Macedonia. Yousef told us he was thinking of going overland through the Balkans, all the way to his fiancée in Sweden.

Just a few days after our arrival in Athens, Omar had rung me up while I was alone in our room, telling me to come to Victoria. "I got a surprise for you," he said, insisting I hurry over. When I got there, I saw a gray-haired, stoop-shouldered figure beside Omar. It was Firouz, our Iranian Kurdish friend from Moria, who'd had bad luck with the Egyptian smuggler. "Salaam, Habib!" he said as we embraced.

Firouz had just arrived in Athens. He explained that he'd simply bought a ferry ticket and walked on, right past the cop who didn't bother to check his ID. He did look rather Greek with his silvery, fluffy eyebrows.

The next day, Firouz asked Omar and me to meet him at a Persian restaurant nearby, one with a winged Zoroastrian *faravahar* on its signboard. We found Firouz on the patio with three others: his adult son, Shahin; and another Iranian father and son, the elder bald with yellow circles under his eyes, the younger with a long beard and tattooed forearms. The three of them had been stuck here for several months already and were living in the makeshift camp at the old airport on the outskirts of Athens. They all wanted to reach Germany, and were trying to find a way out of Greece.

We ordered tea and discussed what we knew. The weather on the Balkan route was getting worse each day, and smugglers' prices were rising. There were at least five borders between Greece and Germany, and getting past them meant sneaking through the woods and mountains. Another option was to go to Italy by sea, if you were willing to try the trucks at the Greek port of Patras. Flying out of Athens cost thousands of euros and, as always, you might go to prison if they caught you with fake papers at the border.

"Before we can even think about leaving, we have to solve another problem," Firouz said. "We don't have documents to be in Athens. If they catch us trying to leave, they might send us back to Moria. We need a *three-leaf.*" That was the Persian nickname for the folding asylum card, an easily forged piece of cardboard with a grainy photo printed on it.

"There's a place where you can get fake documents," Shahin said. "The Pakistani bazaar."

He offered to take us there. Firouz said his leg was hurting, and asked Omar and me to check it out for him. As we rode the subway downtown, Shahin, who was dark haired and beaky, with a troubled gaze, explained that he'd landed with his brother on the Greek islands at the beginning of the year. By then, the Macedonian border was closed to everyone except Syrians, Iraqis, and Afghans. The two brothers had tried to pass as Iraqi Kurds during the nationality screening. Shahin passed his interview, but his brother had been discovered, so he'd decided to stick with him, not realizing how hard it would be to leave Greece once the border closed completely.

"Where's your brother now?" asked Omar.

"Forget him," Shahin said. "He's selling drugs and stolen phones in the camp. He shamed us."

We got off the train at Omonia and rode the escalator up to the square, with its knot of honking traffic. We walked south, past the city hall which faced onto a broad plaza with statues of Theseus and Pericles. At the other end there was a fenced-off archeological site, where an ancient cemetery had been excavated and its bodies exhumed, among them a boy buried with his tortoiseshell lyre some twenty-five centuries ago. We were just outside the walls of ancient Athens. From here the main road had led south to the Agora, center of an empire that had attracted migrants like the philosopher Diogenes, who came from the Black Sea and lived like a stray in a terra-cotta cistern of the sort that had housed refugees during the Peloponnesian War; the Cynic had confounded Athenians by announcing he was a *kosmopolitis*, a citizen of the world. Perhaps more than a quarter-million people lived in Classical Athens at its peak. Plato and Aristotle, who believed history was cyclical, wouldn't have been surprised to see Athens decline to a town of fifteen thousand, a backwater of splendid ruins, by the time of Greece's war of independence from the Ottomans in 1821, but the Orthodox villagers had faith in another city to come, the New Jerusalem. Their patrons, the Great Powers, sent them a Bavarian named Otto as their king; Otto built the plaza we were now crossing and named it for his uncle Ludwig; in 1977, it was renamed the National Resistance Square, to honor the fight against the Nazis, although Athenians still called it Kotzia Square, for the mayor who'd demolished the old Municipal Theater here after it was squatted by ex-Ottoman refugees during the Exchange of Populations of 1922, people who'd burned the theater's furniture to stay warm, and spoke in dialects and foreign tongues, naming Athenian neighborhoods like Nea Smyrni for their lost Asian cities, whose plangent music, *rebetiko*, was today played for tourists in the tavernas above the excavated Agora.

Entering the cramped quarters of downtown Athens, Omar and I followed the three Iranians into a narrow lane where men stood in watchful groups or waited alone, hot goods arrayed on tarps on the sidewalk, phones mostly, the air sharp with masala and betel from carts selling chickpeas and *paan*. It got busier on Sophocles Street, with its long row of shops, the Soulehria Brothers, Raja Jee Fast Food, New

Hong Er Da Import-Export, Dubai Shopping Center, and Shalimar Computers, the lettering in Greek, Arabic, English, and Mandarin. Shahin spoke briefly with a Bangladeshi shopkeeper, who shook his head. We pushed deeper into the bustle, until a radio warbled and Shahin froze.

"Look, those dogs are here," he hissed.

At the intersection, two motorcycle cops were questioning a dark-skinned, older man, his papers in their hands. The crowd was slowly moving away.

"I guess this isn't the right time," Shahin said, and led us around a corner. Just then, though, he bumped into another Iranian, who pulled us into a covered arcade. The man fished out an asylum card and handed it to Omar. I peered over his shoulder; the card, dirty and tattered, had the photo of a chubby Pakistani on it. The guy wanted a hundred euros for it.

Omar shook his head.

IT TURNED OUT THAT FIROUZ had been to Athens before, more than twenty years ago. He'd landed in Greece with the first wave of Kurdish refugees in the nineties, and spent a summer working fruit harvests before making his way to Germany, where his wife and daughter were still living. For reasons he didn't elaborate on, he'd returned to Iran, a decision he'd come to regret.

"Let's go see the lights," Firouz suggested to Omar and me one day, as we were hanging out in Victoria Square.

"The lights?"

"You'll see."

We followed him to a street nearby, where the buildings had glass lanterns outside that were left on during the day and kept burning until late. Firouz stopped in front of a purple door.

"Go on," he said to me. "Go in."

"Why?" I said. "What's in there?"

"What are you, scared?"

I opened the door and, followed by Omar, walked down into a low-ceilinged room with a couch at one end and a bead curtain obscuring the other. The air was humid and perfumed, and a disco light cast a dizzying pink net around us.

"*Kalispera*," called a honeyed voice, and an older woman with an elegant perm swished in. Her expression curdled when she saw Omar and me.

"No sex! Only Greeks!" she squawked in English, and spun on her heel. We stared at each other for a moment, dumbfounded.

"Go!" the madam shouted from behind the curtain.

"Only Greeks, hmm?" said Firouz, as we stumbled back into sunlight. "It didn't use to be that way."

The streets around Victoria, I learned, were part of an informal red-light district that extended downtown, getting cheaper until you reached Metaxourgeio, where migrant customers were welcome and the women sold sex, or were forced to sell sex, for as little as ten euros. Prostitution was legal in Greece, but official licenses were seldom granted and the people themselves were often still illegal, many from African or ex-Soviet countries. Victoria and Alexander parks were used by male prostitutes, a glut since last year of young migrants who traded blowjobs for five euros. I often got propositioned by Greek men: "Afghanistan?" they'd ask hopefully. There was one striking individual who circumambulated Victoria in the evenings, his vast belly preceding him like the prow of an icebreaker—he'd find a youth and vanish but would be back an hour later, seemingly insatiable, eyefucking you as he passed.

Afghanistan stood out for the number of minors, almost all boys, who arrived in Europe without their parents. Some had been separated from older companions on the way, but many set out on their own or in groups of children, typically with the consent and financial support of their families. Sons were part of a diversified survival strategy: one for the farm, one for the government, one for the Taliban, and one for Europe. In some villages, it was understood that once you were sixteen, it was time to get going, because if you made it to Europe before your eighteenth birthday, you were treated with much more compassion. Some countries, like Sweden, even allowed your parents to join you, if only you could make it there.

When the border closed, thousands of unaccompanied children were trapped in Greece, some living on the streets where they mingled with the drug dealers and johns. I met a sixteen-year-old boy who'd been living on his own at the abandoned airport. He told me that one day another young Afghan invited him to a local man's house for dinner. He was hungry and broke, so he went. Afterward, he thanked his host for the meal, and the old Greek said there was no need, but if he was really grateful, then he could come give him a kiss. The man had a drinking glass with a photo of an Afghan teenager printed on it; he'd kept that boy for six months, before giving him money to go to Germany. The sixteen-year-old fled.

These homeless boys had posed a dilemma for the activists at Plaza. Legally, the children were supposed to stay in centers run by the state and NGOs; in reality, the system was so overwhelmed that kids were being held in jails. The boys often ran away from the centers, anyhow, in order to find a smuggler. Most wanted to try the trucks at the port of Patras, because it was cheap and minors didn't have to worry about being fingerprinted in Italy. In the end, as an emergency measure, the squat had hosted a group of teenagers on the seventh floor, a random family that played and fought together and were watched over by the other residents. Zied, the volunteer from Tunisia, had some gloves and shin pads and taught them kickboxing. I watched one session; the boys loved it. "You want to go to Germany?" Zied shouted, pummeling his sparring partner. "There are Nazis there! Fight! Fight!"

They were boys who'd crossed the desert with baby fat on their cheeks, boys who'd outlived older brothers. One of the eldest had apprenticed to a tattoo artist in Iran, and his ink spread across the residents' skin: a butterfly for Jamila, a wolf for Mustafa. Another named Ezat, a pale, serious boy with a lisp, learned enough English to help the other residents during their visits to the hospital. Ezat was born a refugee in Iran, and though he'd had to work in a brick kiln from the age of twelve, he found time to read on Fridays, and developed an obsession with stories about Paris. He liked to find a patch of grass and sit looking down at his book, pretending that, should he lift his head, he'd see the Eiffel Tower. He'd read uninterruptedly for as long as he could to maintain the illusion.

By the time Omar and I showed up at Plaza, the seventh floor was emptying out. Some of the boys joined family members elsewhere in Europe through a legal reunification process; others went into the Greek system, once suitable places opened up.

The rest slipped away to the trucks in Patras.

AT THE END OF OCTOBER, we heard the camp at Calais was being cleared by the French riot police. The Jungle, as they called the shantytown of ten thousand, stood on the shores of the English Channel at the far side of Europe's refugee underground, an antipode to Athens. At night the migrants there tried to sneak onto trucks headed to England; the Afghans called it *andakht*, using a verb which can mean to shoot or throw but here might be translated as "trying"; the year before in the Jungle, a young musician named Abdullah composed this song to a traditional melody:

> *The whole Jungle seems empty*
> *And I miss dear Ibrahim*
> *The whole Jungle seems empty*
> *And I miss dear Ibrahim*
> *Qasim is trying a lot*
> *He knows all the ways*
> *He doesn't sleep in a tent*
> *There is Bola Jan*
> *He gets sick from sleeping too much*
> *And I miss dear Ibrahim a lot*
> *The whole Jungle seems empty*
> *And I miss dear Ibrahim*

Across Europe, what became known as the *long summer of migration* was coming to an end. A million had moved through the continent, but now the networks they'd carved northward were being closed down by the law and cold weather. Those who tried to breach the fences faced harsh violence. Omar, horrified, showed me a video on Facebook of Afghans at the Hungarian border, weeping as they recounted being beaten by guards and forced to crawl through barbed wire.

We stopped seeing our Syrian friend Yousef around the square, and wondered if he'd made it to his beloved in Sweden. Then one day Omar got a set of voice messages from him on WhatsApp, which we asked an Arab resident of Plaza to translate. Yousef was in Serbia, he said. He and another Syrian had gone to northern Greece and paid a Pakistani smuggler two thousand euros each to take them through the Balkans, but the man abandoned them in the mountains of Macedonia. They were lost and it was raining. Yousef said they spent a couple of nights in the bush, with the temperature close to freezing, and might have died had a police patrol not found them. They were locked up for two weeks in a filthy cell and then the Macedonian cops took them to a deserted stretch of the Serbian border, and told them to cross. Now Yousef was stranded in Belgrade, broke and homeless. The city was full of stuck refugees, he warned us, and it was getting colder. The way to Hungary was closed; it was too dangerous to cross. We asked our Arabic translator to text him, offering to help.

Whatever you do, don't come this way, Yousef replied a couple of days later. We never heard from him again.

19

We left Plaza without saying goodbye, like the others had before us. Omar and I had already been there for two weeks, longer than we'd anticipated. Haunted by the prospect of Laila being married off in his absence, Omar was anxious to get to Italy. We donned the secondhand coats we'd scrounged, hoisted our backpacks, and slipped out first thing in the morning, while the rest of the squat was still asleep. They had been busy until late preparing for a celebration to mark Plaza's sixth month in existence, and I was a bit sorry we were going to miss the party—though we left our room key with a friend, just in case we had to come back.

We took an intercity bus west, bound for Patras. After what happened to Yousef, Omar had given up on going overland through the Balkans. The ferries that crossed the Adriatic Sea to Italian ports like Venice and Trieste offered Omar the chance to go straight to his final destination. You had to pay one of the smugglers who monopolized access to the trucks, but it cost only a few hundred euros. On the other hand, crawling under a truck was dangerous. I told Omar that if he was willing, I'd stick with him, but he was still undecided. He'd heard of another, more expensive option called *night cargo* where the smugglers put you in a container outside the port, usually with the connivance of the driver. Omar decided we should go to Patras and ask around in the abandoned complexes where migrants lived, dubbed the

Wood and Panjshiri factories. These were controlled by rival smuggling gangs that often clashed; just a few weeks earlier, an Afghan had
been stabbed to death in the Wood Factory and the cops shut everything down, but word was that the game in Patras was back to normal.

The bus ride took three hours. We crossed into the Peloponnese,
then followed the Gulf of Corinth. Greece's third-largest city, Patras
held a famous carnival each year ahead of Lent. Then, over the summer, the tourists came in droves aboard ferries where they drank wine
on blankets under the stars. But when Omar and I got down at the bus
station, we saw that the travel agencies nearby were shuttered for the
season, and the waterfront felt deserted. The old ferry terminal had
been right here in the center of town, beside the naval base, where the
admiral of the Ionian Fleet would look out his window and see migrants climbing up mooring lines. There'd been a shantytown back
then on the outskirts of Patras, home to nearly a thousand people,
mostly Afghans, but the police had bulldozed it seven years earlier.
When a new ferry terminal was opened farther south, the migrants
had moved to the abandoned factories around it.

Packs on our shoulders, Omar and I walked down the highway
toward the big ships in the distance. After a few miles, we arrived at a
redbrick complex across from the ferry port, its shingled roof crumbling in. A row of men lounged by the gate, their faces and clothes
smudged with grease, while others crouched at the port fence on our
side of the road. Omar and I joined them and looked through the bars:
two men, one carrying a wooden plank, were creeping through the
grass, stalking a line of trucks a hundred yards off.

"Whose travelers are you, brothers?" asked a stocky man in Dari.
He was missing his right eye.

"We just came from Athens, we don't have a smuggler yet," Omar
answered. Just then, the two migrants came running back toward us,
chased by a guard on a scooter. We all hustled across the highway to
a line of trees. When the pair hopped the fence, the scooter peeled off
back to the line of big rigs. As one of the fence jumpers, an older man,
made it to our side, he hunched over and gasped for breath.

"See those trucks?" said the one-eyed smuggler. "You climb underneath them, and that way you'll get past, right onto that boat." He pointed to the massive ferry.

"Isn't it dangerous?" Omar asked.

"No."

An enormous man in sweatpants came over and squeezed our hands. "Whose travelers are you, friends?"

Omar explained that we were new. "Is this the Wood Factory?"

"No, this is the Panjshiri Factory," said the one-eyed man. "The Wood Factory is down there. That's where our dear countrymen the Hazaras are," he said with a sneer.

"So what happens if the police catch you?" Omar asked.

"Under the truck? Nothing. They tell you to get lost. If you're inside the truck, then they put you in jail."

"But they don't beat you?"

"They don't even lay a finger."

A younger kid with a black daub on his cheek interjected: "Those *padar nalat* got me in the balls. I was under a truck and they were poking around with a stick to see if there was anyone there. They jabbed me right in the nuts, and I screamed."

The older jumper was still panting, but his guide came over and tapped him on the shoulder: "Come on, let's go. They're gone."

The two jogged across the road and jumped the fence, the smuggler in one fluid motion, his traveler with some difficulty. But they retreated almost immediately. This time, a guard ran up after them to the fence and stuck his phone through, filming us. We turned our backs or ducked behind the trees.

"He's gonna put it on Facebook," said the grease-daubed kid, snickering. The Greek picked up the plank they'd dropped, and started kicking at it, struggling to break it, which prompted more derisive laughter.

The giant in sweats asked us not to stand in the road with our bags, and invited us inside, but we thanked him and said we'd be back. We headed down the highway, curious to see the Wood Factory.

A half mile down, past a supermarket, we saw another compound with a four-story office building and a red-roofed warehouse, with ABEX in Greek letters on them. We walked along the wooden fence until we came to a spot where the boards had been pried off, and climbed through.

Through a cluster of undergrowth, we could see people standing in a courtyard the size of a soccer field. Saplings and grass pushed up through the cracked pavement. The open-sided warehouse wrapped around the yard; to the right was a hangar, three stories tall, its windows and corrugated tin panels broken, so that we could see into its shadowy depths. To the left was a raised loading dock with three tents pitched on it. A few migrants sat there on scavenged furniture, while others stood in the yard conversing with a group of young men and women in sweaters and jeans, who looked like aid workers.

We approached the loading dock, where there was a firepit surrounded by charred blocks of cedar and sooty metal tins. A jowly middle-aged man rose from his camping chair and greeted us warmly. He introduced himself as Haider, from Kabul.

"You need to find a good smuggler here. We're not happy with ours, Abu Fazl," he said, jerking his chin toward a tall, bearded young man. "Don't pay cash, put it to sleep with a money changer, that way you won't get stuck with one smuggler."

"Who's the best?" Omar asked.

"Jawad Fence is pretty good. That's him over there." He pointed to a man in gray shorts and a knit cap crossing the courtyard.

"What about Rambo?" Omar had heard his name back in Athens; he was supposed to be the best.

"Rambo and Uncle aren't here right now. They'll come tomorrow."

It turned out that Omar and Haider had worked for the same US-AID contractor in Kabul. As they were comparing notes, a younger man drew up to me, wearing a filthy pair of tear-off track pants. We looked each other over: his nose tapered to an upturned point and his cheeks were flushed.

"Salaam," he said. "Where are you from?"

"Kabul."

"Where in Kabul?"

"Shahr-e Nau."

"Where in Shahr-e Nau?"

"Qala-e Fatullah."

"Where in Qala-e Fatullah?"

"Wazirabad Street."

"Really?" he drawled. "I'm from Wazirabad and I know everybody there. How come I've never met you?"

I was flustered but Omar came to my rescue, peppering the kid with his own questions. They dueled over the names of neighborhood police chiefs and high school principals for a bit, until the kid, satisfied of our provenance, stuck out his hand. "I'm Sharif. I've been here in Patras for three months, and don't worry, I'll show you how everything works."

The aid workers came over. They were from Praksis, a Greek medical charity. The local coordinator, a young woman named Maria, was showing some visiting Belgian colleagues around. She greeted us through her translator, an Afghan wearing a button-down shirt and square glasses. Praksis ran a home for unaccompanied minors in Patras, but she explained that they also had a drop-in center where adults could get breakfast and shower.

"I was in the best underage home in Athens, and it was run by Praksis," exclaimed an Iranian kid in a tracksuit.

I heard a squeaking noise and turned to see a man, naked except for a pair of grayish boxer shorts and slippers, pushing a shopping cart holding a Lidl bag stuffed with clothes. No one else paid him any attention as he crossed the courtyard.

Sharif was staring at Maria, his blush deepening. "Ask her if there's anything that they can do for a broken heart," he told the translator.

Maria smiled but didn't answer.

"You know, we were fifteen, fifteen years ago, so maybe you can give us a place in the underage home," Omar joked in English. The Belgians chuckled. Omar started telling them about his work with the coalition forces in Afghanistan.

I wandered off deeper into the warehouse, marveling at multilayered lofts connected by ladders and stairs. Though the building's surfaces were all rusted or splintering, its cedar beams, steel girders, and concrete slabs seemed like they might stand for a century. I walked into a second courtyard, in front of the four-story office building. The shards in its window frames caught the setting sun, and when I looked up past the lines of laundry hanging from the upper floor, I saw a silhouetted figure climbing a ladder to the tower on the roof.

When I returned, Maria and her crew were gone, and Omar was kicking around a soccer ball with Jawad Fence, Sharif, and another Afghan kid. Omar asked the smuggler about going as *night cargo*, as an alternative to trying to crawl under a truck.

"I used to do that, but not anymore," Fence said, in a thick Hazaragi accent. He had a pleasant, weathered face, entirely devoid of hair. "It's expensive. Besides, I have too many travelers right now. Why don't you just try?" He smiled at Omar's hesitation, and turned to the others. "When people come from Athens, they're always afraid at first."

"I was scared too, but now it's natural," said the kid. "I've learned all the places under the truck." There were three spots to hide: the tool box; the spare tires; and the axles, which offered the best concealment but were the most dangerous, since you could get caught in the suspension or driveshaft. "Take a wooden board to put across them," he told us. And we had to remember to crawl between the two rear wheels, so that we wouldn't be crushed if the truck started moving.

Fence shook our hands—his grip was like an iron claw—and strode off toward the office building.

"What about you, Sharif?" I asked. "How come you've been here so long, don't you want to try?"

He laughed heartily. "I try in places so dangerous that no one else dares," he said. "But the cops caught me each time. Now I'm waiting for a friend who's getting out of prison soon, so we can go together. I'm just helping the smugglers for now."

The man with the shopping cart, who'd been bathing and washing his clothes at a water spigot, was squeaking back toward us in his boxers.

"Who is that?" I asked Sharif.

"The black guy? I don't know his name."

"Where is he from?"

"He says he's from Portugal."

"Hello," I said in English as he passed. He stopped and smiled. He was short, and looked middle-aged, though he didn't have a trace of fat on his stocky body. Long dreads hung over his shaved temples. Yes, he was from Portugal, he said. I asked him how he had ended up in this place—wasn't he allowed to get on the ferry and go back home?

He met some people here, he said, and got drunk with them, and woke up in an alley with all his belongings gone, including his passport. He'd been in Patras eight months, and though he'd gone to the Portuguese embassy, for some reason they wouldn't give him another passport. I had trouble following the story—his words were getting more jumbled and rushed as he spoke. "Everything here is a chance," he said, glancing around skittishly. "So many people come, they go, Afghan, Syrian, they sleep here and then they go. But I stay."

He gripped the cart's handles, and then turned back toward me. "I go to the city in my mind, you know what I mean? I need to experience things."

He wheeled his cart onto the loading dock, and past the tents, toward a set of stairs that led up to a loft. A couple of Afghan kids helped him carry the cart up. There was no exterior wall, so you could see him in cross section, as in a dollhouse, stringing up his clothes.

It was getting dark. Haider had the fire going, and was slicing up a bag of raw chicken livers while a sad tune from Naghma played on his phone. Two cats, calico and mackerel-striped, were rustling through the garbage. Another Afghan man who was traveling with Haider, bald with round, professorial glasses, got up for his ablutions, reminding us not to miss our prayers.

"We're at the end of our rope coming here," Haider said. "It's not me that I'm worried about. It's the boy." He was here with his ten-year-old son. They'd been trying to escape Greece for eight months. They had tried twice to cross the Adriatic on small craft with smugglers, but were intercepted both times.

Haider's son walked over and peered at the glistening liver, his face grave under a bowl cut. He looked small for his age. A young woman came up and stood behind him, dressed in a baggy sweater, her chestnut hair uncovered. She watched us silently in the firelight as Haider stroked his son's head.

"We're going to London, God willing. I have family there. I got some French ID cards for us in Athens. I'd show you them but they're sewn into his sweater." He tapped his hand between the boy's shoulder blades. "We'll use them once we get to Italy. With an ID, you can buy train tickets and stay in hotels."

"Why don't you use it to buy a ferry ticket?" Omar asked.

Haider shook his head. "It won't work. They'll catch you at the port in Italy, and fingerprint you."

"Where are you from?" the woman asked, in an Iranian accent.

"Kabul," said Omar. "We just got here."

"We're lucky." She had a high, childish voice.

"Huh? How are we lucky?"

"We're lucky to be here."

Omar snorted. "We are?"

"We're lucky to be here because we'll get past."

"Why are you so sure?"

"Because God is great."

She led Haider's son away, murmuring to him in her singsong pattern.

"Yesterday, she told me her whole story," Haider said. He'd met the woman here in the Wood Factory. She'd grown up in Iran; both her parents were dead. He shook his head. "The poor thing, she's all alone. I told her we'd help her get to Europe."

Haider got up to get some cooking oil from his tent. Omar and I watched the Iranian walking hand in hand with the child. It was the first time on our travels that we'd seen a young woman on her own.

"She came here with an Afghan boyfriend, but he left her behind and went to Italy," Sharif said.

"That baldy gave her three hundred euros," said a curly-haired Iranian teenager who'd joined us. He was referring to Haider's traveling companion, the man with the glasses. "I bet they're fucking her."

Haider returned and started frying the liver with onions over the fire. Hungry, Omar and I went to the supermarket up the road, and, feeling conspicuous under the bright fluorescent lights, bought eggs, tomatoes, onions, pitas, and three tall cans of Heineken. We scurried back and slipped through the fence, two dark shapes slinking into a hole.

Once Haider was finished, Omar put the pan back on the fire, which I stoked with scraps of cedar. He cooked the diced vegetables until they softened, then nestled the eggs among them. When the whites had set, we dug in with hunks of toasted bread, inviting the Iranian kid to join us.

Afterward, we puffed on cigarettes, sated, our fingers gleaming with liver grease. Sharif had gone to see the smugglers and, when he came back, Omar showed him the three Heinekens, which he'd kept hidden in the bag out of respect for the others' religious sensibilities. "Let's go to the park," Sharif said, and giggled. "You can see people kissing there."

We set out up the highway. "Hey, you coming?" Sharif yelled to a solitary figure hunched against the port fence across the road, who waved us off.

"Is he trying?" Omar asked.

"Nah, he's just a junkie."

As we walked along the fence, Sharif explained how the smugglers had carved up the port into territories. Rambo and Uncle had staked out a weighing station across from the Wood Factory, where the trucks had to stop before entering the terminal. "You'll meet Rambo tomorrow," Sharif said. In Dari, the pronunciation sounded like "Rimbaud."

"He must be dangerous, with a name like that," I said.

"Rambo? He's the nicest guy. When you talk to him, you feel like you've known him forever."

At the Panjshiri Factory, there were no migrants by the fence anymore, but we could see the large silhouette of the smuggler we'd met earlier that day, standing by the gate. Sharif cinched his hood around his face. "I'm fighting with them," he whispered. "They're on the outs with us. The time that guy got stabbed, it was them that did it."

The smuggler nodded to us as we passed. When we reached the north end of the port, we saw a white-walled compound behind the fence. "That's the commandos' jail," said Sharif. "They keep you for twenty-four hours. You get a sandwich and a bottle of water."

We cracked the beers as we walked. There was one ferry still at the port, red, with *Grimaldi* on its side. "That leaves late at night," Sharif said.

Suddenly, he slipped his beer into his pocket. "Hide your drinks."

Ahead, we saw Haider and his companions, also out for a stroll. "Are you ready to go tonight?" Sharif asked them, as we drew near. He pointed to the ship. "You want to go on that?"

"Sure we do," said Haider.

"It's fifty-three hours; think he can stand it?"

"The boy?" Haider laughed. "He walked forty hours through the desert, and eighteen hours through the mountains. Sitting in a container is nothing. He's seasoned. He's become a man." He clapped his hand on the shoulder of the little child, who smiled shyly.

"The only thing we're worried about," said the bald man with glasses, "is what to do about the bathroom."

We left them and continued to a seaside park, where there was a group of Afghan boys sitting in a circle by the swings. They were residents of Praksis's underage home. One of them gave a joyful shout, and bounded toward us: it was Ali, the kid with bleached-blond tips we'd met swimming on our very first visit to the port in Lesbos.

"I'm a *khod andaz*," he whispered in my ear as we embraced. A self-smuggler. He was trying the trucks on his own, hoping Rambo and the others wouldn't catch him. He'd get a beating if they did; the smugglers were ruthless about their monopoly. "You have to go once it's dark. There's a way to get around the fence by the water, I can show you. We almost got through last night but the police spotted us."

We finished the beers in the park and then went back to the Wood Factory. The smuggler Abu Fazl was there, telling Haider and the others to get their stuff ready. They grabbed their bags.

"Say a prayer for us," Haider asked.

"Don't pray, don't say goodbye, just go *qalandari*," Sharif retorted—like a wandering mystic. They shook our hands anyway.

Omar and I had only brought sleeping bags, but Sharif said there was a tent we could use up in the office on stilts at the back of the hangar. We climbed up a ladder; it was like a treehouse, neatly finished in cedar. There were people already sleeping on the floor, and several domed tents. Omar and I found the empty one and crawled in.

"What do you think?" I asked him, as we lay side by side in the dark. He answered in English.

"I think it's fucking dangerous, bro."

IN THE MORNING, WE FOUND Haider and his crew sitting around the ashes of the firepit, looking glum. Abu Fazl had put them inside a container, but it was full of scrap metal and there was no place to hide among the cargo. The cops had discovered them in the inspection bay, where the trailers were opened and searched. But they were all let go without being arrested, likely because of the woman and child.

"I was in a container full of liquor once, but there was no place to hide, so I got down," the curly-haired Iranian kid said. That was the luck of the draw. You might end up in some stinking recycling, or a load of roasted pistachios.

After a quick breakfast, Omar and I went to the weighing station across the road. A dozen migrants were lined up at the fence, crouching behind its concrete base. On the other side, there were two lanes of trucks waiting for their turn to get weighed, about a dozen yards away.

"Where are Uncle and Rambo?" Omar asked them.

"Over there," an Afghan kid answered, pointing to two squat men in track pants and hoodies. The smugglers were taking two or three clients at a time over to the other side of the fence, where they lay in the grass, shielded by a slight rise, and waited until a truck stopped on the scale. Then they ran toward it and tried to stow their travelers aboard before it drove off, inside if the container was unlocked, underneath if not. All this took place in full view of the drivers farther back, who sometimes sounded their horns in alarm. They rarely got down to roust the migrants, preferring to leave it to the guards at the inspection bays.

I heard Omar exclaim and turned to see two of the teenagers from Plaza: Reza and Ezat, the boy who'd dreamed of Paris. They'd just arrived on the bus from Athens. And they knew me as Matthieu, not Habib. As we hugged, I tried to think of how to warn them, before they blew my cover in front of the smugglers—but just then, Rambo shouted for them to come over.

The boys hopped the fence as a truck was pulling up. They sprinted to it, and Uncle and Rambo each grabbed a handle on the container and wrenched the doors open. As they did so, the truck started forward and they jogged with it as the boys scrambled inside. Now working at a dead run, the smugglers slammed the doors shut and the truck sped off to the inspection bays.

"Those lucky *khar kos!*" exclaimed one of the migrants. "They aren't here an hour and they get a good one."

I watched the trailer drive off toward the inspection bays with the boys inside, feeling worried for them but relieved my secret was safe. Rambo and the other might not react kindly if they learned I was a journalist. The trucks were coming less frequently now. The smugglers managed to put one more client inside a truck's chromed toolbox before a cop screeched up in a black BMW and we all ran for it.

BACK AT THE WOOD FACTORY, Rambo and Uncle were sprawled out in the sun, rolling cigarettes. Under a faded ball cap, Rambo's blunt, craggy face was clean-shaven, exposing a long scar on his upper lip. Uncle looked a little older, in his forties maybe, short but with the same vise-grip handshake as Jawad Fence.

I asked Rambo what kind of container Ezat and Reza had gone in.

"We put them in a truck carrying flowers," he said. "I don't put my travelers inside bad cargo. They're underage, after all."

The two smugglers had been working in Patras for more than a decade. Back when the ferry terminal had been located in the city, it was monopolized by Kurdish gangs. But more and more Afghans started showing up, and a war broke out between the two groups, with the

Afghans led by a legendary *badmash* named Patras Khan. Eventually the Kurds gave up and went north to another port, Igoumenitsa.

Rambo and Uncle retired to the office building; there wouldn't be any more trucks until later that afternoon. Omar was listening to Sharif pour out his heart about a woman he'd met in the camps.

"Sharif, if you love a girl, don't let her go," Omar said.

I lay back on the warm pavement and thought of the two boys in the dark amid the flowers. Once inside the steel box you were trapped until someone opened it for you. The fear of suffocation must be intense, especially the first time. Perhaps the two were already sailing to Italy, the passengers above unaware of the children below. Did the boys feel terror, or elation? They were trying to reach Germany—Ezat would take care of Reza until they got there, I knew. Ezat was a serious boy, and focused. He'd been spending long hours at a workshop I'd visited in Exarchia, a little one-room store with a handpainted sign: ATELIER MOHAJIR. Inside were three heavy-duty sewing machines, strong enough to sew through thick canvas. I'd go there sometimes to escape the chaos of the squat, and sit with Ruby, the Egyptian German woman who ran the workshop. She looked younger than her thirty years, but Ruby was motherly to the teenagers at Plaza, and gentle with me until I mentioned the men I'd seen picking up boys in Victoria. Then her gold-green eyes flashed. "I'd like to stab one of them," she said through gritted teeth.

When Ruby was still a fashion student in Berlin, she'd fallen for a local while vacationing in Morocco; the fact that she could visit him, and not vice versa, had seemed fundamentally unjust. She got involved with the Movement and No Borders activism and, during a visit to Lesbos several years ago, met a group of Afghan boys who'd landed in Europe alone. She was struck by their plight and realized that while the kids would accept food and shelter from the NGOs in Greece, eventually they'd escape back to the underground, with all its dangers, to continue their journey. What the boys needed most of all was the money to reach their destination country up north, where they'd be safe. Ruby wanted to do something to keep them away from the johns and dealers on the streets of Athens. She knew a lot of the kids had worked in sweatshops in Iran and Turkey, and were able tailors; in Germany, bags made from recycled truck tarpaulins were popular, and she got the idea to reuse the dinghies that filled the dumps on the islands. She raised money to cover her own expenses so that the price of each item that she sold at festivals and street fairs in Germany— canvas backpacks, shoulder bags, and tobacco pouches for ten to sixty euros—went directly to the refugee who'd made it. She wasn't anyone's boss; she gave each of the tailors a key to the atelier, so that they could come and go as they pleased. She stuck a piece of tape with their name to each item, so she'd know whom to pay, and her customers often asked her not to remove it. Sometimes they wanted her to pass on a message or photo to the kid.

Just a few years ago, Ruby told me, not many people in Germany were aware that refugees crossed the sea in such flimsy craft; now some found the choice of material macabre. What if someone had died in that boat? they asked. It was unlikely, she'd reply, since the boat had made it ashore, but yes, it was possible. People drowned trying to reach Europe. Did they ever think about the stories behind the phone in their pocket, or the jeans they wore?

Handling the canvas was emotional for some of the tailors at first, when the trauma of crossing was still fresh. But habit rendered the substance inert, as they reworked it to their own design. The basement of the workshop was filled with deflated dinghies taken from Lesbos. I climbed down once and sifted through the pile, feeling the different weights, the varying strengths of seams, until I found a piece in the same green and gray as the boat that took me and Omar across.

AT DUSK, EZAT AND REZA returned to the Wood Factory, covered in mud. They'd tried to hide between the crates of flowers, but just before the inspection bay, the Panjshiris opened the container and inserted one of their own clients. Inside the port, the commandos stripped the tarpaulins off and found Ezat and Reza. They suspected there was one more migrant, but they couldn't find the guy—he'd crawled deep beneath the muck of the crates, and they didn't want to unload all the cargo. So he'd made it onto the ferry. The officers let the boys go, since they were underage.

Relieved that he still hadn't blown my cover, I pulled Ezat aside. "People here know me as Habib, from Afghanistan, OK?"

"OK." He nodded, his expression serious. He was a smart kid. He didn't ask me what I was doing there, just like I wasn't going to convince him to stop trying. What could I do, threaten to call the cops? Ezat was here for the game, and would keep playing until he made it. Already, he was numbing himself to the danger. Children, agile and brave, were well suited for Patras. It was the adults who lost heart. I didn't think Omar was going to try.

When night fell, we went hunting for scrap wood for the fire. The walls and floors had been stripped of boards, but working in concert we dragged over a cedar beam and stuck one end in the fire. For illumination, we gathered fat sheaves of invoices from the office, yellow and white carbon copies, each page giving a burst of light before it curled into smoke. Whole forests must have passed through here, I thought, looking at the figures listed in drachmas.

As we warmed our hands, Omar and the others talked of the good lives that awaited them in Europe, the jobs and homes that would make the price they paid worthwhile—if only they could go *pesh*, forward.

"Greece is ruined like this factory," said Haider, gesturing toward the maw of the hangar, where drifts of garbage and clothing lay like dark snowbanks, layers left by those gone before us, awaiting those to come, the people walking now in deserts and mountains, through tunnels and highway medians, fleeing war and poverty, and believing, like my companions, in progress, in moving toward the place where I stood, faithless—yet it was for them that history had been a *catastrophe which keeps piling wreckage upon wreckage*, as another refugee, Walter Benjamin, once wrote.

THE NEXT MORNING, OMAR TOLD me he wanted to go back to Athens. He had another idea, a better one than throwing himself under a truck, but we had to return to Plaza first. That was fine by me. We grabbed our packs and said goodbye around the campfire. The foul-mouthed Iranian kid surprised me with a big blind hug, like I was an older brother.

Just as we were about to leave, I heard a squeaking sound and turned. Maria appeared, leading a group of Praksis staff and student volunteers, who were pushing a cart with rubber tires.

"We've come to play games," she announced in English with a bright smile. They were doing outreach aimed at unaccompanied minors, and as they unfolded the cart, I saw it was the sort of thing you'd bring to a child's birthday party: panels with spinning wheels, a map, a chalkboard, and a magnetic pad where you assembled animal shapes.

It was the adults, though, who seemed most interested in playing with Maria and her young volunteers. Sharif smoothed his hair and stripped off his track pants, revealing cleaner jeans underneath. He and others, including Omar, got in a circle with Maria, who started off with a name game. I watched for a bit, then noticed Ezat standing off to the side. "Don't you like games?" I teased him. He bit his lip.

"It pisses me off. They come here and go, 'Oh, how horrible, you poor things,' and take pictures and then they show them to their boss and he says, 'Good job, here's your salary,'" Ezat said. "I mean, I know Praksis is giving real help as well, and I'm grateful for that. I just don't like these people that come with their hearts bleeding for us."

The players were doing a relay race now, but Ezat's gaze drifted into the distance, as if seeing his own road ahead—his twelve attempts, the container of diapers in which he'd make it onto the ferry, the forty-eight hours he'd spend trapped inside, the church in Italy where he would take refuge from the police, the train ride without a ticket to France, the freezing alleys of Paris, the Champ de Mars where he would stand trembling in ecstasy, Hamburg where he would be granted asylum, where after two years he'd learn enough German to start university, his past as inscrutable to his classmates as his future would be to his family in Iran, living alone in body and mind, the cold of the River Elbe in winter seeping into his bones.

BACK IN ATHENS, NO ONE at Plaza had noticed our absence. Omar and I went up to our room and talked his plan over. We had both left our documents at the squat when we went to Patras, in case we ended up on a truck to Italy. I had forgotten he had an ace in the hole: Vygaudas. In theory, the Lithuanian ID would allow him to travel out of Greece; as long as he stayed within the Schengen zone, he didn't need a passport. But there were police at the airport on the lookout for escaping migrants, and if they stopped him, he would have to pass as Vygaudas. Visiting Patras had given him the idea to take the ferry from there to Italy, as a passenger—surely no one at the port spoke Lithuanian.

Omar bought a ticket to Trieste and changed back into his black outfit, the one that had brought him luck on Lesbos. I would let him go ahead alone, and if he made it, I'd follow using my own passport. It was a twenty-six-hour cruise up the Adriatic and once Omar set foot on Italian soil, he could surrender himself and claim asylum. Our journey would at last be at an end.

The ferry left in the evening. Omar took the bus back to Patras and caught a cab at the station, which took him down the highway, past the Panjshiri Factory and the migrants at the fence line. His ship was alongside, dwarfing the terminal. The clerk at the travel agency had told him to show up two hours before departure, but boarding hadn't started yet. Omar sat down in the lounge, one of the few passengers that time of year. The guards were watching him; he took out *Report to Greco*, and opened it on his lap.

Without biometrics, catching someone with fake or stolen documents can require old-fashioned skills. Border guards have only a limited amount of time to spend on each passenger; in a few seconds, they take stock of your appearance and body language, and the way you respond to a question. A lifetime can be determined in a heartbeat. *If the hand of the Afghan man had not shaken, Hamid would now be living in Norway and not in Canada*, wrote Shahram Khosravi about a fellow refugee busted in Delhi. *Border crossing is, after all, a matter of performance.*

It was time to board. The guard took Omar's ticket and ID, looked at his face, then back at his documents. He told Omar to wait. A cop came over: "Follow me, please." Omar went with the man to a side room, and he told himself not to panic. Eventually an older officer arrived and started talking to him in a foreign language he didn't understand. Omar stared back blankly.

"What, you didn't learn Russian in school?" the cop asked in English.

They made him stand and he felt the cold metal click around his wrists. They rifled through his bag and took him to a holding room, where he saw the boy Reza from the factory, slumped in a chair.

"What happened?" Omar asked in Dari, dropping any pretense of being Lithuanian.

"They caught me on top of a truck," Reza said. "I think Ezat got through."

There was a heavily pregnant Afghan woman in the room as well; she was quietly weeping.

"Why are you crying?" Omar asked.

"Because our money's all gone," the woman answered. She and her husband had already been caught at the airport. They'd spent the last of their savings on ferry tickets and fake documents and then split up, thinking it would improve their chances. She had gone first. A policewoman came over and asked Omar, in English, what the matter was.

"She's scared," he said.

"Tell her not to worry, we'll let her go tonight," said the officer. "We're good people, we won't hurt you."

"My friend, where were you last year?" her partner quipped. "The borders were open then."

Reza and the pregnant woman were released without charges, but Omar was transferred to jail, to face a judge in the morning. So this was it, he thought. Even if he didn't get a prison sentence, they might figure out he'd escaped Moria, and send him back to the island.

His cell had two bunks and a toilet in the open. A fair-haired boy in his twenties was already there when Omar arrived. He was from

Georgia, in the Caucasus, and said he'd come to Greece with his mother and sister several years ago, for a job on the island of Santorini, at a fancy restaurant where some bottles of wine cost sixty euros.

Omar let out a low whistle. "It sounds like you had a good job," he said. "Why did you leave?"

"I wanted to see London," said the boy. He'd tried to board the ferry, but his visa had expired. He was afraid of what would happen next. "If we get out, come to Santorini, I'll find you a job there."

In the morning, they were both taken to court, where two blond women waved from the gallery—the Georgian's mother and sister.

The Greek judge was younger than Omar expected, in his thirties maybe. Omar went first. Feigning ignorance, he told the judge through the interpreter that he'd been tricked by the smuggler, and thought the ID card would let him travel legally.

"You broke the law," said the judge.

"I'm sorry. We're not criminals, but we have no choice. We're fleeing war in our countries. People from Afghanistan, Iraq, and Syria didn't come to Greece for sightseeing." Omar looked at the judge. "If you were in my place, you would do the same thing."

"Maybe, I don't know," the man said, seeming amused. "Why don't you stay in Greece?"

"There is no work. You're a judge and you have a good job, but many of the people in your country are jobless. How would I find work?"

The judge told Omar he was letting him off with a warning, but if he was caught again, he'd go to jail.

The cops led Omar through a hall and opened a door. He stepped out into the sunshine.

20

I woke with a start, my heart sprinting until I reached up and found the wooden frame of the hotel bed. In Moria, rising from a nightmare meant waking to something worse. I looked left and saw the mattress beside me was empty. Omar must have gone to get breakfast. I turned my head to the right: it was bright outside. I'd slept in. As I lay back, a vague sense of dread came over me. It felt like I was forgetting something.

I'd spent the night of Omar's arrest waiting in our room, my phone in hand. I was ready to call a lawyer the next day but he'd returned to Athens, free but downcast. We'd been so close to ending the journey. It was hard to believe that a year had passed since I'd flown back to Kabul, thinking we were about to leave for Europe.

I could hear kids playing in the hallway. At least Omar was safe here in Plaza. I didn't mind spending time in the squat, with all its distractions. The six-month celebration was going to start tomorrow. Various working groups had been busy planning a three-day conference with panels and workshops on the No Borders Movement, capped by a dinner and concert, all open to the public. Today, the squat was supposed to mobilize for a deep clean. We'd scrub out the fridge and storerooms, shift couches to uncover each moldy cucumber and contraband beer bottle. The preparations were so thorough that a rumor had spread among the residents that a celebrity would visit Plaza; though who exactly, no one could say.

I sat up—November 9; wasn't the US election today? I tried to figure the time difference. We were seven, or was it eight hours ahead of New York? The polls wouldn't open until late in the afternoon in Europe; the results would come in the early hours of the morning. There were no TVs in the squat, and I'd hardly been paying attention to the news. The last time I'd looked, the *New York Times* gave Hillary Clinton an 85 percent chance of winning. Maybe they'd updated their forecast. I reached for my phone.

While I was asleep, Donald Trump had been elected president.

"IS THAT THE FAMOUS PERSON?" the Afghans at my table kept asking each time a likely foreigner spoke to the crowd jammed into Plaza's dining hall. The conference sessions were being held in Greek and English but there were tables with Arabic and Persian translators, and I'd volunteered at the latter. The latest speaker my tablemates were asking about was tanned and had thick glasses: Sandro Mezzadra, professor at the University of Bologna, whose research described how borders had become *mobile, permeable, and discontinuous*, lines woven into society, which crossed all of us—probably not whom my table had in mind. But how would we know a celebrity when we saw one?

For three days, visitors from around Europe passed under the watchful eyes of the Antifa Club and on into the mezzanine, where they browsed the portraits and posters and cooed over the children who were silenced by the influx of strangers for once. In its short lifetime, the squat had become famous among the European left, and the conference had attracted the committed and curious alike. "I have never before seen a space like this!" exclaimed a man in a beret to his companion. If each chair was occupied and people sat along the cupboards in back, at least a hundred of us could fit into the dining room, our space of assembly, *la plaza*.

The activists living here wanted to connect the squat to the big picture, but the news the panelists brought was bad: as in America, anti-immigrant populism was on the rise across Europe, a backlash to the million who'd entered during the Long Summer. The dream of open borders is a nightmare for many. And yet the activists, too, wanted to end mass migration—only not with walls, but by ending war and dispossession. That meant the world system based on nationalism and capitalism had to change; instead, it seemed stronger than ever.

Zied, the volunteer from Tunisia, gave a presentation on how Europe's border crisis had moved to the central Mediterranean. On the route between Italy and North Africa, a far more dangerous passage than the Aegean, upwards of five thousand had drowned in 2016—a record number of deaths, mostly African, which provoked little of last year's outcry.

"SO WHY DID TRUMP WIN, if Hillary got millions more votes?" Kalliroi asked me. I was working the nighttime security shift with her and her friend, both from a leftist student group that supported Plaza. You spent the shifts at a table by the entrance, and filled the hours with cigarettes and idle talk, but you had to stay alert in case the fascists came and threw a Molotov through the window, as they'd done a couple of months earlier at the squat on Notara, where refugee families were living. In a corner behind the table, there was a bundle of pickax handles with strips of red, black, and purple fabric tied to them, so they counted as flags at demonstrations—Marxist red, anarchist black, and queer purple, the latter at the students' insistence.

I tried to explain the peculiar institution of the Electoral College, and why the Founding Fathers were so worried about majority rule. Kalliroi shook her head scornfully: "Anyway, red team, blue team, it's the same ruling class."

Nasim came in and interrupted: "Please tell the residents to be very careful when they go outside," he said, looking even more tired than usual. "The streets will be full of police. People should have their papers on them."

The most famous man in the world was coming to Athens, and it was going to be messy. Obama's final tour overseas as president had coincided with the anniversary of the 1973 student uprising in Greece against the military dictatorship. This was the most important demonstration of the year for the Greek left, who'd never forgiven America for supporting the junta, and every November 17, a giant crowd assembled at Polytechnic University and marched on the US embassy. Obama's visit, the first by an American president in seventeen years, would add fuel to the fire.

For several days around the anniversary, the university campus hosted a street fair, and one afternoon, Plaza's activists corralled as many refugees and volunteers as they could find. Around forty of us walked over to pay our respects, with Omar and me bringing up the rear. "They ruined it!" gasped a little Afghan boy as we entered the campus, his eyes wide at the dense churn of graffiti and tattered banners that covered the buildings. His mother hushed him; Nasim laughed. The yard was filled with tables and posters whose acronymic logorrhea denoted parties and student groups tracing their lineage to the original Greek Communist Party, the KKE, founded in 1918. The tabletops were piled with materialist journals and pamphlets. I recognized a slim paperback, *The Communist Manifesto*; nearby was a translation of Naomi Klein's book on global warming, *This Changes Everything*.

A decade ago, the leftist party currently in government would have been scattered among these marginal factions—the name Syriza stood for Coalition of the Radical Left. In the 2007 election, Syriza had won only 5 percent of the vote. Back then, Greece was still dominated by two increasingly similar parties run by the scions of political dynasties—on the eve of the debt crisis in 2009, the center-left candidate, whose father and grandfather had been prime minister before him, assured voters that *there's plenty of money*. But the old consensus was shattered by the economic collapse and the punitive austerity imposed afterward by Germany and the other creditors, which drove Greece even further into debt. New forces emerged, like the Golden Dawn, whose blackshirts blamed Jewish bankers and which briefly became the third-largest party in Parliament, and Syriza, led by the young and charismatic Alexis Tsipras, who, soon after he won office, had reneged on his promises to end austerity and the mass detention of migrants, enraging the activists who'd supported him.

Even the children grew somber when we arrived at the memorial, with its twisted metal gate, crushed by a tank on November 17, 1973. Beside it lay the enormous bronze head of a youth, resting on a slab as if decapitated. The chairs around us were occupied by boys in leather jackets, KKE activists whose pale faces reminded me of photographs of martyrs. Over the microphone, someone announced that *Refugee Accommodation and Solidarity Space City Plaza* had come to pay its respects to the memory of the uprising; Olga, an activist at the squat whose parents had taken part in the struggle, handed out carnations, which we laid at the broken gate.

For security reasons, the Syriza government had banned demonstrations downtown during Obama's visit, but Athens's radicals were going to march anyway. The day of the president's arrival, a couple dozen people from the squat met at Steki, the social center in Exarchia. Elias, a young Greek Web designer who helped run the maintenance team at Plaza, stood before us, his wavy hair tied back in a ponytail. I liked Elias for his calmly ironic manner, but he was grave now as he addressed the group in English. "The government is serious about blocking the march from reaching the embassy, and the entire city is full of security forces," he said, looking at each of us in turn. "There is a chance of mass arrests. Violence is guaranteed."

Omar and I both wanted to join the march, but after hearing Elias speak, I told Omar it might not be a good idea for him.

"Then why are you still going?" he asked.

"I want to see it," I said. Besides, I still had my passport in my pocket.

"OK, brother," he replied, giving me a searching look. "Stay safe."

Polytechnic University wasn't far from Steki. When we arrived, Nasim and Elias were surprised at how many people were gathered there already, awaiting the start of the march. The crowd stretched for blocks, clustered by party under hand-painted banners and flags. We found space at the rear, between some Greek Maoists and a crew of Syrian boys from one of the squats—in Athens, angry young Muslims became leftists. While we waited, I asked about the latest version of the celebrity rumor I'd heard: that Obama wanted to visit Plaza. Elias laughed.

"Obama's people asked someone at Human Rights Watch for suggestions for his itinerary, and that person asked us if we were interested," he said. "Of course it was impossible." Both the Greek government and the US embassy would have surely vetoed it, but the request annoyed the activists—as if they'd ever let the Secret Service inside Plaza.

"People see news stories about us, and think we're some kind of NGO," grumbled Nasim.

"Does it feel strange to protest against Obama, now that Trump has been elected?" I asked Elias, as the crowd started moving. "Things might get worse for refugees."

He shook his head. "Even if he is more progressive than Bush, or now Trump, he's still the representative of the main imperialist power that's responsible for the wars that made many of these people refugees," he said.

I recalled an anarchist poster I'd seen at the university: *The visit of the US President Obama on the day of the Polytechnic revolt of 1973 is a provocation that won't remain unanswered.* Obama had wanted to speak near the Parthenon, symbol of democracy, but the event was moved to a more secure location after a grenade was thrown at the French embassy. Threats had been made by the Conspiracy of the Cells of Fire, self-styled nihilists. So instead, the president would give his speech at the Stavros Niarchos Foundation, named for the Greek billionaire. It was a swan song, of sorts—it had been eight years, I realized, since I'd stood in the shower in Herat listening to his speech in Chicago, when he told us that a change was going to come.

After walking for half an hour, we arrived at the first row of riot cops lining the sidewalk in helmets and gas masks, farm boys from the Peloponnese, where they still held God and country dear. Behind every dozen shields stalked an officer armed with a big can of pepper spray or a tear-gas launcher. The march moved forward in stops and starts through downtown Athens. It was overcast and getting colder, the light drained of color on the concrete and metal shutters and the shabby clothing of the marchers. Only the red of the flags stood out. The chants were getting angrier now, calling Tsipras an American lap-dog, denouncing *planetary terrorism*, and in them was the cadence of protests past, the dead generations. Behind us, the Syrians were sing-ing a raucous call and response in Arabic against the dictator, waving the flag of the revolution with its three stars.

"Pigs! Cops! Killers!" A column of black-clad anarchists pushed up through the marchers. They wore gas masks and motorcycle helmets, and gripped pickax handles with flags. Some carried fire extinguishers or backpacks heavy with projectiles. The Black Bloc wedged itself be-tween our group and the Syrians and formed into a square, the outer ranks linking arms, their clubs held horizontal. Nasim and Elias ex-changed a look of dismay. There was only one way this would end.

We'd reached a choke point ahead of Syntagma Square, where the cops were letting groups through one at a time. The Black Bloc and the riot police were skirmishing, taking runs at each other. Elias was on his phone, getting updates from ahead. There were some scuffles and tear gas, nothing serious, but now it was the anarchists' turn to pass. A dumpster flared brightly with a Molotov, forcing the cops back.

A helicopter roared overhead. I saw the photographers had donned their masks and were moving in for the shots that would make the evening news.

A chant in English went up: "Refugees and immigrants welcome! No to planetary terrorism!"

I opened my mouth, but nothing came out.

There was a volley of blasts, and grenades came bouncing into the crowd, spraying fumes. Our chef Cristian dashed up ahead amongst the Black Bloc, and swung his red flag inside the cloud.

I want to commend Prime Minister Tsipras, Obama would say the next day, *for the very difficult reforms his government is pursuing to put the economy on a firmer footing.*

The riot cops charged forward; our ranks broke and we fled back down the boulevard. Flash-bangs went off around us and I nearly hit the deck. Elias was yelling to slow down, stick together. There was more fighting back the way we came. I saw our banner duck up a side street and followed at a jog.

At the top of the alley, our group assembled in an empty plaza. Our retreat had been fast and ignominious and we were mostly intact, although the gas was starting to burn, an ache that reached down through your eyes until your whole body throbbed. Cristian arrived, his face wet and beet-colored, and one of the Greeks squirted him with a spray bottle full of antacid.

Catching my breath, I looked around and realized where we were: the enormous university square where, two years ago on my first trip to Athens, I'd watched Tsipras give his victory speech. *The verdict of the Greek people ends, beyond any doubt, the vicious circle of austerity in our country*, he'd told the jubilant audience. For a week, I'd barely slept in the apartment I'd rented with friends in Exarchia, up each night in smoky bars hearing about the revolution to come. It wasn't just Syriza but Spain and Portugal, Bernie Sanders and Jeremy Corbyn. To believe felt like falling in love.

As I looked around the deserted square, blinking back tears from the gas, I saw again the joyous faces that had filled the streets that night, heard their mellifluous laughter. After the victory speech, we'd gone to the Syriza tent nearby, where the delirious crowd had wept and embraced, linked arms and sang:

> *So comrades, come rally,*
> *And the last fight let us face.*
> *The Internationale*
> *Unites the human race.*

AFTER BREAKING UP THE MARCH, the cops chased the anarchists back to Exarchia, and fought running battles around the square. A few of us from the squat went there later and watched the mob pile chairs into tangled barricades. They uprooted the steel posts that lined the sidewalks and used them as pile drivers to smash the paving into projectiles. A crosswalk light swung from its cord like a piñata, until someone brought it down with a baton to cheers. The cops stood two blocks away behind their shields, and threw rocks back at us, their tear gas answered by a Molotov's rainbow frozen in midair.

Those were the last days of the autonomous zone. Syriza would lose the next general election to the conservatives. Greece's new leader, an investment banker whose father had been prime minister, vowed to clean up Exarchia, and promised he would not allow *a revival of a new generation of terrorists.* The cordon of riot police grew tighter, until their boots stood on the square.

And Exarchia would become Athens's second-most popular neighborhood on Airbnb.

SO WHAT IF THEY DOUSE the candles in rooms where lovers meet? wrote the poet Faiz. *If they're so mighty, let them snuff out the moon.*

On the last night of Plaza's six-month celebration, the squat had held a dinner and dance, open to all. Omar and I were on the kitchen shift; Shero, the Kurdish chef, prepared roast chicken and rice. To make the occasion more formal, we served people in the dining room, carrying our trays over the children playing between the tables. Shero's pièce de résistance was homemade ice cream, flavored with cardamom and brought out to great acclaim. After the hectic service, Omar and I sat with the rest of the shift on the little kitchen balcony and passed around a steel mixing bowl of leftover ice cream, pressing the metal against one another's sweaty foreheads. The moon was nearly full.

By the time we finished cleaning, the band had set up at the end of the dining hall cleared of tables and chairs. Olga, the activist, came to the microphone; the residents kept cheering for her each time she tried to speak, until the joke got old and she fixed us with her famous no-nonsense stare, which redoubled our laughter. When we'd finally subsided, she introduced the band, a Kurdish group, and explained they'd been invited out of solidarity with a leftist party in Turkey whose leadership had been arrested. It was just a gesture in the face of state repression, but it had brought us together that night. Maybe that's what no borders meant; you could start in your heart, with your neighbors, in your city.

Backed by a drum machine, the singer, guitarist, and keyboard player struck up party music, the kind you'd hear at a Kurdish wedding, the singer exhorting the crowd as the guitar skittered and the synth threw loops. The Kurds formed a circle, men and women linking arms, and danced clockwise in a three-step: two steps right, one kick left. Atop that rhythm, a whole variety of movements can arise, and as the crowd gets bigger, concentric circles rotate in opposite directions, while in the center pocket people take turns kicking and twirling, a scarf pulled taut between upraised fingers.

The Kurds started things off their way, but the rest of us joined and it mutated into a freestyle. An Afghan resident, carrying his little son, led his group inside; the *atan* dance was not so different. Old Hajji from Iran busted out what were either traditional Khuzistani moves or just his personal repertoire, back hunched and jazz hands aflutter. Zied hoisted Rabi from Syria on his shoulders, so I put Carles on mine, and we added, to wolf whistles, a vertical layer to the rings.

If we only understood love, Kierkegaard said, we would not need to be commanded. The truth we sought at Plaza had not yet come into this world; perhaps we'll only find it in fragments. Yet we strove toward union in that moment, spinning round and round together, the songs getting sweatier. As I came past, I saw Omar standing in the outermost ring with the hand-clappers and phone-filmers, his arms crossed. I caught his gaze and called his name, but he just watched, the smile remaining on his lips through the revolutions.

21

As winter set in, Omar despaired of ever escaping Greece. I overheard him on the phone to Maryam in Istanbul, lamenting the danger of going under the trucks, the risk of using fake documents at the airport. The judge in Patras had let him off, but he had a police record now, and next time he might not be so lucky.

He was spending more of his days in bed, with his Samsung held to his nose, as if he were going to push it through the screen. I'd bought him the smartphone for our trip; he'd never had one before, and now he fell hard for Facebook. Each time I came to the room there was some grotesquerie he insisted on showing me, a viral clip that proved djinns were real, or a woman being trampled and set alight by a mob in Guatemala. His relatives in Afghanistan posted about suicide bombings; his friends who'd made it to the West wrote about how they were lonely and sick of working dead-end jobs. The one person Omar wanted to see wasn't on Facebook at all. Laila's father still hadn't returned her cellphone, and Omar hadn't spoken to her since he left Kabul. All he had was street sightings from neighbors. The meaning of his odyssey depended on her: whether he'd possess a wife, a family, and a home. Or else he'd end up wandering a wilderness, alone.

Increasingly, I found him listening to Céline:

Near, far, wherever you are
I believe that the heart does go on

Each time I came in and saw Omar lying there, his face in Facebook, I felt a prick of annoyance. What kind of protagonist was he? Nasim asked me why Omar didn't get more involved in the squat; the Afghans needed their own equivalent of Rabi, the Syrian who helped manage the front desk, someone who spoke English and understood Europe, who marched with the activists and made love to volunteers, a real hero.

Omar and I started bickering over silly things, like cleaning the room. When he asked for money for more phone data, I made a gibe about Facebook; we didn't speak the rest of the day.

I KEPT MYSELF BUSY WITH tasks like chopping onions in the kitchen or wiring the windows shut in the stairwell, after I caught some kids trying to climb out on the fifth floor. I liked the aleatory flow of communal life. I could make myself useful as a translator for the doctors who visited the squat. With female patients, I sat on a stool facing the corner, my voice a pain inside, a failing heart, a history of cancer, of psychosomatic seizures and self-harm. For some of Plaza's dreamers, it was a thousand and one of the same night, the plane overhead, the dumpster slamming in the alley. But the war was over in Europe.

We need courage to face life, as well as death. Zied, hagridden, had to quit translating for the Syrians and the psychologist. He wanted his sleep back; he told me he was thinking about flying home to Tunisia for a rest. Five years before we met, Zied had been in the sea of thousands outside the interior ministry in his city of Tunis, when the people's roar of *bread, freedom, dignity* had shaken the earth. In a couple more years, my friend Zied would volunteer for the rescue ships off the coast of Libya, helping to pull migrants from the water, their bodies covered in gasoline burns, men who paid their passage in desert gold mines, women who took contraceptive injections before setting out. What is homecoming when a day's flight can take you back?

OMAR AND I COULD NEVER stay mad at each other. We were lying in our bunks, idly discussing the squat's latest romances, when he let out an abrupt laugh. "Can you believe it's been a year since I've been with a woman? I used to need it every week, at least, or I'd go crazy." He got up, paced to the balcony door, and lit a cigarette. "How is it possible that I've changed so much, all of a sudden?"

I told him that it was fine, even to his advantage, if he didn't need sex so badly. But I had expected him to be more excited by Athens's charms, if not by the women, then by the atmosphere of freedom.

"It's because I love her," he said.

I reflected for a moment. Was that the explanation? I'd kept my doubts about Laila to myself, not wanting to interfere, but sometimes when he asked about life in the West, I'd dropped hints about how much easier it was to be independent, and about how long and costly his scramble upward would be. I don't know if he listened; I might have been talking to myself. Now I was frank. How could he be so sure he loved her? He hadn't spent much time in her presence. They hadn't even walked openly on the street, let alone passed a night naked in each other's arms. They knew each other over the phone. Perhaps he only loved his idea of her. Perhaps he loved a fantasy.

He stayed quiet, looking out the window, and I went on. His desire to marry Laila had come at the same moment that the rest of his life had fallen apart, as his friends and family were leaving Kabul for good, and the American visa revealed itself as a mirage. Could he have been grasping for something else to give him purpose? "I just don't want you to sacrifice so much for something that might not be real," I said. He turned toward me.

"It's true that she's my only goal now," he answered, softly. "And I can't stop thinking about her. I'm always calling her friends and asking about her, even though I know it makes me look like a donkey, someone to be pitied."

He didn't recognize himself depressed like this. He'd once been filled with a lust to live, so much that he'd frittered away the surge years on parties and love affairs, while his brothers secured their futures in Europe.

"Who am I? What happened to me?" He was looking at himself in the mirror now. "Look at me! Who are you?" His reflection gave a sad laugh.

The light outside had dimmed. Through the wall, we could hear our Syrian neighbors arguing. Omar sighed. "Even if coming to Europe turns out to have been a mistake, if I'm ruined here or I go back to Kabul, I'll never forget how you helped me. You're a true friend. You helped me with a pure heart."

I looked down at the carpet. "Just get to wherever you need to go, and then we're even, OK? I don't need any thanks. Just find yourself again and be happy. That's all I want."

"I have to change, brother. I want to change."

"I know. You will once you get where you're going."

WHEN I PROPOSED TO WRITE about our journey, more than a year ago, neither of us knew how it was going to end. Sure, Omar wanted to get to Italy, but fate might have easily taken him elsewhere. We didn't expect the trip to last so long, though; I imagined I'd have time to stay awhile with Omar once he reached his city of refuge; maybe get a job under the table while he applied for asylum, living down and out together, him from necessity and me of my own free will. I'd even fantasized about applying myself. I could begin anew as Habib, live years as him, a lifetime. Could a body still do that? Or would Matthieu's fingerprints cross me one day and bring the police to my door? We were entering a new era of biometric surveillance, where the undercover journey I'd made would become impossible.

We all have things about ourselves we'd like to change, and it's seductive to imagine it happening in one swift movement. That was the dream behind migration: a fresh start. The journey was a prelude. Life came afterward, and it might be harder, more heartbreaking than the smuggler's road.

But in truth, we can't leave ourselves behind. We get only one story, which we narrate looking backward. Our choices and chance encounters, the trembling of another's hand, all matter because of where they lead us. Alasdair MacIntyre defined us as a *story-telling animal*; our endings give us meaning.

And yet those ends circle back to swallow themselves. If Omar's conclusion to this trip was a beginning, I, too, would have to restart the journey, living it again until it was language, and the person I'd become, a character on the page.

"After this, you won't come back?" Ruby asked me. "You won't be part of the Movement?" We were sitting alone on the hill above Exarchia, talking about the community that had formed around Plaza, and whether it would last. Ruby, like the others, knew I was writing a book, but I'd confessed to her the secret of my escape from Lesbos, and told her that I still felt like an impostor inside the squat, although not because of my false identity.

"I don't know what I'll do," I answered. She looked at me with compassion, saying nothing, her slender arms clasped around her knees. Ruby, full of love for strangers' children, would become a mother in a few years. As for me, I was going back to New York.

No trucks, Omar decided. He would try the airport with fake documents. The smugglers' prices had come down since their peak that summer. But if he was going to fly, then he could go anywhere—that was the magic of airports. Did he really want to end up in Italy, where he heard that work was nearly as hard to find as in Greece? He wanted to learn a trade, so that he could support a family. If he got accepted in a rich country like Germany or Sweden, he'd get financial assistance to study. The asylum process would take time but there was no guarantee that Italy would be faster. He had to have patience, and trust Laila to wait. This way, he would have more to offer her family.

Whatever longing subtracts from the heart, wrote Rumi, *surely it will bring better in exchange.*

If Omar was going to fly, then it wasn't a good idea to follow him through the airport. And there was no telling how long he'd have to wait for a smuggler to send him. Christmas was approaching, so we came to an agreement; I would return home for the holiday, and if Omar was still stuck in Athens after the new year, I'd come back. Otherwise, I'd see him in his new country, wherever that was.

As it happened, Firouz, our Iranian friend, made it out of Greece first. He flew with a fake passport to Germany, where his wife and daughter were waiting. His success heartened Omar, who then found his own smuggler, a cocky young Afghan who'd grown up in Greece. The man met with Omar in the back of a Chinese store near Victoria. His price was four thousand euros, placed in escrow, and Omar could keep trying until he made it—unless, of course, he went to jail. When the smuggler saw how European Omar looked, he laughed and slapped his shoulder. "You'll get past, a hundred percent," he said. "If you don't make it, I'll give up this work."

I PACKED MY FEW BELONGINGS and said goodbye to Nasim, Zied, and the others. On the way downstairs, I stopped in the mezzanine under the row of portraits and tried to fix the scene in memory. The squat had become a haven for many kinds of displaced, a foothold in the city against the centrifugal forces of incarceration, deportation, gentrification, everything that was forcing people out to the end of the train lines, to the government camps and drab suburbs. But Plaza wasn't permanent, we all knew that. Even now, we were scattering. We find each other just to lose each other.

Omar was waiting for me on the sidewalk under the signboard. Years ago, I'd seen him at the Mustafa Hotel, a stranger. Now there were tears in his eyes. We embraced and squeezed tight; then I had to go.

———

THE *HAWALA* NETWORK IN ATHENS was connected to the one in Istanbul, so Maryam could pay the smuggler there. Omar called his mother and explained where to take the money.

"I'm going to fly as soon as the smuggler is ready, Mother."

"May God protect you."

Maryam knew her son risked going to prison, but all she could do was pray. Her faith was strong. Hadn't God protected all six of her children through the war, when so many mothers had lost their own? Maryam had carried Omar as a newborn into the camp in Pakistan, and seen the hillside lined with little graves. Now he was a man in Europe. Their lot in life had been poverty and violence, but she had refused to accept it. They were Afghans, but they were also human beings. Although survival had meant separation, Maryam believed they would all be reunited one day. She and the others would follow her son, even if they had to cross the water.

I HAD TO GO BACK to Trieste to collect the laptop and bags that I'd left there earlier that summer, so I decided to take the ferry. It was an overnight journey to Italy from Patras but I wanted time to think. I bought my ticket downtown, not far from the ancient cemetery at Kotzia Square. From the fence along the archeological site, you could look down into the open graves. This was a yard for commoners; the skeletons from another site near Kerameikos had richer diets. Many of the people here must have been slaves, who weren't buried separately. Perhaps a third of the population of Classical Athens was enslaved; they were integral to the city-state's economy. *For that some should rule and others be ruled is a thing, not only necessary, but expedient,* wrote Aristotle; *from the hour of their birth, some are marked out for subjection, others for rule.*

Only a minority, the adult male citizens, could vote in the assembly. Just as women's work at home gave men free time for public life, slavery provided the material basis for Athenian democracy; the philosopher Bernard Williams argued that ancient Greeks were unable to imagine civilization without it. *The effect of the necessity was, rather,*

he wrote, *that life proceeded on the basis of slavery and left no space,
effectively, for the question of its justice to be raised.*

WHEN HIS DAY CAME, OMAR went to the airport. The smuggler met
with his clients one at a time in the parking lot and gave them their
documents. Omar got a Bulgarian passport with his photograph in
it, and a boarding pass. There were twelve people, including him, and
they were all booked on the same flight to Switzerland. From there,
they'd connect to their separate destinations. Rather unusually, their
smuggler was flying with them, accompanied by his girlfriend, a styl-
ish young Greek. It was almost Christmas, and the two were going on
a Swiss holiday.

This time, Omar was careful to wait for a while after the others
went inside. He smoked in the parking lot, pacing between cars, thinking about his previous chances in Lesbos and Patras—his luck, good
and bad. When he was ready, he entered the terminal, and joined the
flow moving toward the security lines.

IN PATRAS, I THOUGHT I recognized some of the smugglers in front of
the Panjshiri Factory as my taxi drove past. We pulled off the highway
and into the port, where I paid the driver and shouldered my backpack.
The ferry terminal was almost deserted and as I walked toward the
boarding gate I could tell the cop and security guard were sizing me up.
I guess I should have been more careful, dressed as I was, looking like I
did, but I didn't have the heart for the script: purposeful stride, eye contact, arrogant smile. I just fished my documents out of my pocket and
handed them over. The cop took the passport, flipped through it, and
then went into the office. Through the doorway, I could see him pecking
at a computer, maybe checking the Interpol list of stolen documents. A
second guard moved behind me in case I ran for it. Involuntarily, a scene
played in my mind's eye: I would turn, kick him in the groin, sidestep,
and sprint for the exit; I'd have a head start, I would make it to the
parking lot and probably over the fence.

———

OMAR PUT HIS BAG IN the scanner, along with his belt, phone, wallet, lighter, cigarettes, and chewing gum, and then walked through the metal detector. He waited. There was a problem with his bag. The woman was asking if it was his. He shrugged. He didn't speak English. He was from Bulgaria. She rummaged through it and pulled out a can of hairspray, apologizing as she confiscated it. She said he could go. He put his stuff in his pockets, grabbed his bag, and kept moving through the airport.

I SLIPPED MY PASSPORT BACK into its leather case and walked up the gangway, smelling the creosote in the dock, hearing the lines creak as they came under tension. I threaded my way through the ship's bulkheads, went up the stairs and past a mirrored saloon, climbing until I reached the open deck. From the stern rail I could watch the trucks coming aboard the ferry, their trailers banging as they rolled up the ramp. At the far end of the lot, I noticed some figures by the port fence, watching. Each container looked the same, a steel box.

OMAR LOOKED AT THE NUMBER above the gate, and then at the people sitting there, careful not to linger on any one face. He counted eight clients, including himself, who'd made it through. Thirty-two thousand euros. It was indeed Christmas for the smuggler. Omar took a seat. When the passengers started boarding, he lined up near the back; the others were already through when the flight attendant scanned his paper, looked at his passport, and frowned.

"Sir, you have someone else's boarding pass," she said.

I STAYED AT THE RAILING even after the white city receded, watching the wake spiral out, the churn fold on itself. The ferry felt as steady as an island.

"Do you think it will ever get all *gadwad*, mixed-up, like that again?" Maryam had said to me once in Istanbul. She was asking whether the border would open. Would the jubilee come again? I told her I didn't know.

Maryam had become a refugee almost forty years ago, and yet Afghanistan was still at war. In the future, her grandchildren would tell her story to their own children here, to Europeans. But if Maryam's tale inspired because of the long odds that she had survived, then it was also a testament to the many who had vanished. In this way, our stories carry forward fragments of others, just as we pass on our siblings' genes, though they be childless. *We are a synthesis between the conquerors and conquered*, Gloria Anzaldúa wrote; *a blending that proves that all blood is intricately woven together.* We are all the kin of migrants.

Far astern, the coastline ended and a band of horizon opened to the south, aglow with the last light. Over that curve lay the path of our journey, the bombed cities and hungry villages, the desert and mountains, the sea like ink, the islands where children clung to the bellies of trucks. Night was falling on the underground now, Maryam was saying her *maghrib* prayers, Malik worked late in the tailor shop, and Zulmay and Raja lay down in their tent in Moria. Others were setting out into darkness, as we had.

A COLD FEELING SPREAD IN Omar's chest; the smuggler had given him the wrong boarding pass. He blinked, opened his palms in helplessness. The woman told him to wait while the other passengers boarded. Two police officers arrived.

"Bulgaria, Bulgaria, no English," Omar said. He handed them the passport and smiled.

WHATEVER SEPARATES US, WE TRAVEL on the surface of a sphere. I know our paths must cross.

———

THE COP GAVE OMAR BACK the passport and nodded to the agent. Her machine spat out another boarding pass. She waved him through to the gate.

God made them blind.

THE SEATBELT LIGHT CHIMED. A couple of hours passed.

There were mountains below, snowy spires steeper than the ones back home. This was Switzerland; Omar recognized it from the Bollywood films, with their lovers' duets set in dales. But instead of Abhishek Bachchan and Karisma Kapoor, he and Laila, singing the song he would sing to her one day:

See the signs of the changing season
Alone like this, I cannot live

He loved her. I didn't understand, but he forgave me. It was his life, after all. He was going to find a country and make a home. He would learn the language, be grateful but not lose his pride, and when he had earned his right to travel he would return for her.

Was he acting out of necessity or freedom? One did not exclude the other. For what else is life's journey but a search for our beloved?

Epilogue

City Plaza closed voluntarily in 2019, after the election of a conservative government in Greece.

In September 2020, the camp at Moria burned to the ground. The Greek government began building a new, more permanent facility.

As of May 2021, thirteen thousand asylum seekers were held on the Greek islands.

On August 15, 2021, the Taliban captured Kabul, sending a new wave of Afghan refugees into exile.

Maryam, Farah, Suleyman, and Jamal crossed in a rubber raft from Turkey to Greece, and were eventually granted asylum in another country.

Laila and Omar are married and live in Europe with their first child.

Acknowledgments

This book is dedicated to Omar and his family, who shared their lives with me.

In a sense, I've been working on this since I crossed the Amu Darya in 2008. Since then, I have been helped by an array of people too numerous to list here, in particular my friends in Afghanistan, Pakistan, Syria, Iraq, Libya, and Yemen. Some of them are no longer alive.

Over the course of this project, which began in the fall of 2015, I benefited from the kindness and solidarity of many people, a few of whom I cannot name. I am especially grateful to the following:

Rozina Ali, Noah Arjomand, Awista Ayub, Sayed Rahman Bekur, Peter Bergen, Victor Blue, Peter Bouckaert, Bette Dam, Berit Ebert, Fabrizio Foschini, Anand Gopal, Taya Grobow, Susan Kamil, Sarah Leonard, Melanie Locay, Nasim Lomani, Nick McDonell, Shaheryar Mirza, Fazal Rahman Muzhary, Samuel Nicholson, Ilya Poskonin, Zarka Radoja, Siavash Rahbari, Graeme Smith, Katerina Tsapopoulou, Matthaios Tsimitakis, Waheed Wafa, and Thomas Wide.

I wish to acknowledge the following institutions, and the generosity of their supporters, for enabling me to devote five nearly uninterrupted years to writing this book: Type Media Center (formerly the Nation Institute) and the Lannan Foundation, which also hosted me at a residency in Marfa; New America, the Eric and Wendy Schmidt Foundation, Southern New Hampshire University; the Edward R.

Murrow Fellowship at the Council on Foreign Relations; the Berlin Prize at the American Academy in Berlin; and the American Library in Paris. I am also indebted to the Frederick Lewis Allen Room program at the New York Public Library.

I am furthermore grateful to my editor, Noah Eaker, and the publishing team at Harper, to Jacques Testard and Fitzcarraldo Editions in the UK, to Edward Orloff of McCormick Literary, and to Kyle Paoletta, for his work fact-checking the manuscript.

Finally, I thank my own family for their unwavering love and support.

A Note on Sources and Methods

In writing this book, I've stuck to what John Hersey called "the legend on the license"—none of this was made up. The only details I've changed are the pseudonyms I've used to protect people, beginning with Omar and his family.

We left Kabul for Nimroz on August 29, 2016. My principal source for our journey was the more than sixty thousand words of notes I took with my smartphone—an almost daily task I found easier than expected, since everyone's so busy with their phones these days—along with photos and videos I recorded and collected from other people. We finished our journey in late 2016 and, over the next four years, I conducted follow-up interviews and made numerous reporting trips back to the places we'd been. I provided these materials to a fact-checker.

I've also relied on the work of experts in fields like history, anthropology, and migration studies, as well as reports from journalists and other contemporary sources. I've cited references in the endnotes section.

Pseudonym and Chapter of First Appearance

Ali	15
Arash	15
Elham	7
Farah	3

Pseudonym and Chapter of First Appearance

Firouz	13
Habib	6
Haider	19
Haniya	3
Haris	9
Ismail	5
Jamal	4
Karim	7
Khalid	3
Laila	1
Malik	6
Mansoor	3
Maryam	3
Omar	1
Raja	16
Reza	15
Sardar	12
Shahin	18
Sharif	19
Shireen	9
Suleyman	4
Vygaudas	16
Waleed	17
Yousef	16
Zakaria	6
Zia	3
Zulmay	16

Notes

Translations from the Persian and French are the author's unless noted.

1

4   Thousands of people: "Sea Arrivals Data, Greece," Operational Data Portal, United Nations High Commissioner for Refugees (UNHCR).

6   the Special Immigrant Visa: Emmarie Huetteman, "'They Will Kill Us': Afghan Translators Plead for Delayed U.S. Visas," *New York Times*, August 9, 2016.

9   *The wall is also inside each one of us*: John Berger, *Hold Everything Dear: Dispatches on Survival and Resistance* (New York: Vintage, 2008), 94.

2

12   A shocking incident: Matthieu Aikins, "Doctors with Enemies," *New York Times Magazine*, May 17, 2016.

12   a profile of Colonel Abdul Raziq: Matthieu Aikins, "The Master of Spin Boldak," *Harper's*, December 2009.

13   six times what an ordinary Afghan soldier made: Department of Defense, *Report on Progress toward Security and Stability in Afghanistan*, June 2008, 20.

14   another five killed: Ruhullah Khapalwak and Carlotta Gall, "Taliban Kill Afghan Interpreters Working for U.S. and Its Allies," *New York Times*, July 4, 2006.

15   the Canadians launched an offensive: Graeme Smith, *The Dogs Are Eating Them Now: Our War in Afghanistan* (Toronto: Knopf Canada, 2013), 55.

19   I decided to pretend: I wrote about the journey in "Unembedded in Afghanistan," *The Coast*, October 1, 2009.

20   *And to all those watching*: National Public Radio, "Transcript of Barack Obama's Victory Speech," November 5, 2008.

21   nearly all of the world's illicit opium: United Nations Office on Drugs and Crime, *The Global Afghan Opium Trade: A Threat Assessment*, July 2011.

22   bearing signs of torture: I wrote about Raziq's killings in "Our Man in Kandahar," *Atlantic*, November 2011.

22   the number of US troops in Afghanistan would triple: Congressional Research Service, *Department of Defense Contractor and Troop Levels in Afghanistan and Iraq: 2007–2020*, February 22, 2021, 7.

22   "This is a war we have to win": Mark Landler, "The Afghan War and the Evolution of Obama," *New York Times*, January 1, 2017.

3

26   Bette was a Dutch freelancer: Bette Dam, *A Man and a Motorcycle: How Hamid Karzai Came to Power* (Utrecht: Ipso Facto, 2014).

27   a force double the size of the Soviets': Congressional Research Service, Department of Defense Contractor and Troop Levels in Afghanistan and Iraq: 2007-2020, February 22, 2021; Artemy M. Kalinovsky, *A Long Goodbye: The Soviet Withdrawal from Afghanistan* (Cambridge: Harvard University Press, 2011), 42.

27   *Employ money as a weapon system*: David Petraeus, "Multinational Force–Iraq Commander's Counterinsurgency Guidance," *Military Review*, September–October 2008, 3.

27   half a trillion dollars: Amy Belasco, "The Cost of Iraq, Afghanistan, and Other Global War on Terror Operations Since 9/11," Congressional Research Service, December 8, 2014, 19.

27   *third country nationals*: Sarah Stillman, "The Invisible Army," *New Yorker*, May 30, 2011.

30   *This heart came into life*: Hafez Shirazi, "Ay Padesha Khuban," Ghazal 493.

30   Human beauty could reflect divine beauty: Rumi, *The Quatrains of*

Rumi: Complete Translation with Persian Text, trans. Ibrahim W. Gamard and A. G. Rawan Farhadi (San Rafael, California: Sufi Dari Books, 2008), 657, 670.

30   *the You of her eyes*: Martin Buber, *I and Thou*, trans. Walter Kaufmann (New York: Charles Scribner's Sons, 1970), 154.

31   *All across the city*: Rumi, "Man Mast to Diwana," Ghazal 2309, *Divan-e Shams*.

31   Wine washed away the stains of hypocrisy: Franklin Lewis, "Hafez and Rendi," *Encyclopedia Iranica*, vol. XI, fasc. 5, 483–91.

32   *All you who've gone to hajj*: Rumi, "Ay Qawm ba Haj Rafta," Ghazal 648, *Divan-e Shams*.

33   *The world changes*: "Mausam Ki Tarah," from the film *Jaanwar*.

35   More than half of global wealth: Credit Suisse, "Global Wealth Report 2019," October 2019.

35   per capita American income is thirty times that of Afghans: Purchasing power parity, from World Bank data.

35   *citizenship premium*: Branko Milanovic, *Capitalism, Alone: The Future of the System That Rules the World* (Cambridge: Belknap, 2019), 129.

37   not so much brain drain but wealth drain: Gabriel Zucman, *The Hidden Wealth of Nations: The Scourge of Tax Havens*, trans. Teresa Lavender Fagan (London: University of Chicago Press, 2015), 53; Al Jazeera Investigative Unit, "Exclusive: Cyprus Sold Passports to 'Politically Exposed Persons,'" Al Jazeera, August 28, 2020.

37   citizenship, in the twenty-first century, is for sale: Atossa Araxia Abrahamian, *The Cosmopolites: The Coming of the Global Citizen* (New York: Columbia Global Reports, 2015); Transparency International and Global Witness, "European Getaway: Inside the Murky World of Golden Visas," September 1, 2018.

37   *people without anchorage*: Frantz Fanon, *Les damnés de la terre* (Paris: François Maspero, 1970), 150.

4

41   three hundred thousand Afghans were displaced: Internal Displacement Monitoring Centre, *Global Internal Displacement Database*.

41   *catch-22 for refugees*: David Scott FitzGerald, *Refuge beyond Reach: How Rich Democracies Repel Asylum Seekers* (Oxford: Oxford University Press, 2019), 10, 21, 164.

41   one of the worst passports in the world: The worst, according to the 2020 Henley Passport Index.

44   *Life will sing in the end*: Ahmad Zahir, "Zendagi Akher Sarayad."

44   *True Son of the People*: David Edwards, *Before Taliban: Genealogies of the Afghan Jihad* (Berkeley: University of California Press, 2002), 32.

44   Like the Americans in Vietnam: The Russian General Staff, *The Soviet-Afghan War: How a Superpower Fought and Lost*, eds. Lester W. Grau and Michael A. Cress (Lawrence: University Press of Kansas, 2002), 29; Louis A. Wiesner, *Victims and Survivors: Displaced Persons and Other War Victims in Viet-Nam, 1954–1975* (Westport: Greenwood Press, 1988), 349, 355.

45   Afghans had to register: Rüdiger Schöch, "Afghan Refugees in Pakistan During the 1980s: Cold War Politics and Registration Practice," *New Issues in Refugee Research*, Research Paper No. 157, June 2008.

45   *Inside these camps was a huge reservoir*: Muhammad Yousaf and Mark Adken, *Afghanistan: The Bear Trap: The Defeat of a Superpower* (Philadelphia: Casemate, 1992), 138.

45   more than six million people would flee: UNHCR, *The State of the World's Refugees, 2000: Fifty Years of Humanitarian Action* (Oxford: Oxford University Press, 2000), 119.

45   delegates gathered in Geneva: Gil Loescher. *The UNHCR and World Politics* (New York: Oxford University Press, 2001), 45.

46   criteria tailored to the Cold War dissident: Loescher, *The UNHCR and World Politics*, 44, 64.

46   the West could be generous: Gil Loescher and John Scanlan, *Calculated Kindness: Refugees and America's Half-Open Door, 1945 to the Present* (New York: Free Press, 1986), 85, 171.

46   140,000 Vietnamese allies: Figures on Indochina crisis from UNHCR, *State of the World's Refugees, 2000*, chapters 4 and 6.

46   pirates raped and murdered thousands: W. Courtland Robinson, *Terms of Refuge: The Indochinese Exodus and the International Response* (Zed Books, 1998), 43, 60.

46 A new humanitarian politics: Samuel Moyn, *The Last Utopia: Human Rights in History* (Cambridge: Belknap, 2010).

46 *It's very simple*: "Auch einen Zuhälter retten," *Der Spiegel*, October 18, 1981.

47 *at once the temptation to kill*: Emmanuel Levinas, "Peace and Proximity," *Basic Philosophical Writings*, eds. Adriaan T. Peperzak et al. (Bloomington: Indiana University Press, 1996), 167.

47 In his notorious novel: Jean Raspail, *Le camp des saints* (Paris: Éditions Robert Laffont, 1973), 280, 343, 107, 111, 214.

48 Ninety people were taken aboard: Claro Cortes, "Vietnamese Refugees Allowed Ashore after Resettlement Pact with Canada," Associated Press, June 23, 1990.

48 a newspaper photographer took a picture: Deborah Lehman, "Waiting's the Hardest Part," *Times-Colonist* (Victoria, British Columbia), November 30, 1984.

48 fifty-five million Europeans: Thomas Hatton and Jeffrey Williamson, *The Age of Mass Migration: Causes and Economic Impact* (New York: Oxford University Press, 1998), 7.

48 The act was passed after decades of lobbying: Gary Y. Okihiro, *The Columbia Guide to Asian American History* (New York: Columbia University Press, 2001), 111.

48 *Anyone who has traveled*: Greg Robinson, *By Order of the President: FDR and the Internment of Japanese Americans* (Cambridge: Harvard University Press, 2003), 40–41.

49 *They are always going to be brown men*: Alice Yang Murray, *Historical Memories of the Japanese American Internment and the Struggle for Redress* (Stanford: Stanford University Press, 2008), 76.

49 My grandmother Sei: I am grateful to my mother and her siblings for providing family history in this chapter, particularly Louise Anderson, who has collected documents and oral history for decades.

49 whitewashed horse stalls: John Hersey, "Commentary," in *Manzanar*, eds. John Armor and Peter Wright (New York: Times Books, 1988), 7, 70.

49 an inland archipelago: Greg Robinson, *A Tragedy of Democracy: Japanese Confinement in North America* (New York: Columbia University Press, 2010), 154–57, 192.

49   *What has happened lately*: C. K. Doreski, "'Kin in Some Way': The *Chicago Defender* Reads the Japanese Internment, 1942–1945," in *The Black Press: New Literary and Historical Essays*, ed. Todd Vogel (New Brunswick: Rutgers University Press, 2001), 173.

49   chick-sexing: James McWilliams, "The Lucrative Art of Chicken Sexing," *Pacific Standard*, December 14, 2017; Ryan Masaaki Yokota, "Japanese American Chick Sexers in Chicago," *Discover Nikkei*, September 28, 2016.

50   the government's questionnaire: Cherstin Lyon, "Loyalty Questionnaire," *Densho Encyclopedia*.

51   *the fourth enlargement*: Nell Irvin Painter, *The History of White People* (New York: Norton, 2011), 389.

51   Tila Tequila: Amy Kaufman, "The Reality of Tila Tequila's Situation," *Los Angeles Times*, March 14, 2010; Asawin Suebsaeng, "White Nationalists and Nazi-Saluting Tila Tequila Toast 'Emperor Trump' in Washington, DC," *Daily Beast*, November 19, 2016.

51   Unlike Pakistan: UNHCR, *The State of the World's Refugees, 2000*, 116, 119.

53   The end of the Cold War: Loescher, *The UNHCR and World Politics*, 7.

53   *Terrible wars*: Bernard-Henri Lévy, *Réflexions sur la guerre, le mal et la fin de l'histoire* (Paris: Bernard Grasset, 2001), 28.

53   *The world has no business*: "Tribal Turmoil," *Times*, April 27, 1992.

53   *Love died, devotion died*: translated by Wali Ahmadi, *Modern Persian Literature in Afghanistan: Anomalous Visions of History and Form* (New York: Routledge, 2008), 121.

54   *Our commitment*: White House, "President Bush Statement with Afghanistan's President Karzai," September 12, 2002.

54   the first song: Jeffrey Ressner, "Farhad Darya," *Time*, November 26, 2001.

5

56   *She told me that since death*: Elyas Alavi, "Marg dar inja chun gird dar hawast," *Etemad*, 24 Abaan, 1396 (November 15, 2017), no. 3955.

59   The climate was less arid: Matthew Savage et al., "Socio-Economic Im-

pacts of Climate Change in Afghanistan," Stockholm Environment Institute, December 2009.

59   a grand vision for a new Kabul: Matthieu Aikins, "Kabubble," *Harper's*, February 2013.

60   brick factories worked by rural migrants: Samuel Hall, "Buried in Bricks: A Rapid Assessment of Bonded Labour in Brick Kilns in Afghanistan," International Labour Organization, 2011.

6

68   you could bicycle through the Arctic: Peter Tinti and Tuesday Reitano, *Migrant, Refugee, Smuggler, Saviour* (Oxford: Oxford University Press, 2017), 83.

68   At the Democrats' convention: ABC News, "FULL TEXT: Khizr Khan's Speech to the 2016 Democratic National Convention," August 1, 2016.

68   the Afghan militia that the Iranian government was recruiting: Mohsen Hamidi, "The Two Faces of the Fatemiyun (I): Revisiting the Male Fighters," Afghanistan Analysts Network, July 8, 2019.

70   *wide-area persistent surveillance*: Annie Jacobsen, "Palantir's God's-Eye View of Afghanistan," *Wired*, January 20, 2021.

70   MYSTIC program: Loek Essers, "Assange Names Country Targeted by NSA's MYSTIC Mass Phone Tapping Program," *PC World*, May 23, 2014; Ryan Devereaux, Glenn Greenwald, and Laura Poitras, "The NSA Is Recording Every Cell Phone Call in the Bahamas," *The Intercept*, May 19, 2014.

71   American surplus uniforms: Louis Dupree, *Afghanistan* (Princeton: Princeton University Press, 1973), 241.

71   95 percent of Americans' clothing: Elizabeth L. Cline, *Overdressed: The Shockingly High Cost of Cheap Fashion* (New York: Portfolio/Penguin, 2012), 5, 41.

72   *Presto! in the twinkling of an eye*: Jack London, *The People of the Abyss* (New York: Archer House, 1963), 23, 19.

73   *The young American writer*: "Books of Travel and Social Study," *American Monthly Review of Reviews* 28, no. 6, July–December (New York: Review of Reviews, 1903), 762.

73   *going Harun al-Rashid*: Seth Koven, *Slumming: Sexual and Social Poli-
      tics in Victorian London* (Princeton: Princeton University Press, 2006),
      61.

73   *Hating and despising Europeans*: Richard Francis Burton, *Personal Nar-
      rative of a Pilgrimage to El-Medinah and Meccah, Volume 1* (London:
      Tylston and Edwards, 1893), 111.

73   he'd wanted to disguise his Shia origins: Nikki R. Keddie, *Sayyid Jamal
      ad-Din "al-Afghani": A Political Biography* (Berkeley: University of Cal-
      ifornia Press, 1972), 10, 78.

73   *I watched them carefully*: Ali Akbari, *The Illegal Journeys: From East to
      West* (Bloomington: Xlibris, 2011), 91.

74   *Passing, in this sense*: Asad Haider, *Mistaken Identity: Race and Class
      in the Age of Trump* (New York: Verso, 2018), 79.

                                        7

77   *the slave traders of the twenty-first century*: Matteo Renzi, "Helping the
      Migrants Is Everyone's Duty," *New York Times*, April 22, 2015.

77   *Honestly, the people who help*: Dawood Amiri, *Confessions of a People-
      Smuggler* (Brunswick, Victoria, Australia: Scribe, 2014), 109.

78   Iran tightened the border: Alessandro Monsutti, *War and Migration:
      Social Networks and Economic Strategies of the Hazaras of Afghanistan*
      (London: Routledge, 2005), 160; David Mansfield, "Catapults, Pickups
      and Tankers: Cross-Border Production and Trade and How It Shapes
      the Political Economy of the Borderland of Nimroz," Afghanistan Re-
      search and Evaluation Unit, September 2020.

79   The smuggling gangs: Luke Mogelson, "The Scariest Little Corner
      of the World," *New York Times Magazine*, October 12, 2012.

79   the police could be brutal: Human Rights Watch, "Unwelcome Guests:
      Iran's Violation of Afghan Refugee and Migrant Rights," November 2013.

79   Some Americans who were arrested: Shane Bauer, Josh Fattal, and
      Sarah Shourd, "How We Survived Two Years of Hell As Hostages in
      Tehran," *Mother Jones*, March/April 2014.

82   Seventy-three people had been killed: BBC News, "Afghanistan Fuel
      Tanker Crash Kills 73 in Ghazni Province," May 8, 2016.

83 who founded a city here: P. M. Fraser, *Cities of Alexander the Great* (Oxford: Clarendon, 1996), 132.

83 *A hundred and fifty thousand persons were captured*: Louis Dupree, *Afghanistan* (Princeton: Princeton University Press, 1973), 286.

84 Abdul Raziq: United Nations, "Treatment of Conflict-Related Detainees in Afghan Custody: One Year On," January 2013, 49; Rob Crilly, "US General Criticised over Photo-Op with Afghan Cop Accused of Human Rights Abuses," *Telegraph*, February 20, 2014.

84 more than doubled since the year 2000: UNODC data.

84 increasingly solar powered: David Mansfield, *A State Built on Sand: How Opium Undermined Afghanistan* (London: Hurst, 2016), 271.

84 Afghan farmers started to supply: James Tharin Bradford, *Poppies, Politics, and Power: Afghanistan and the Global History of Drugs and Diplomacy* (Ithaca: Cornell University Press, 2019), 131, 177, 187, 208.

85 in the old American-built canal zone: Matthieu Aikins, "Afghanistan: The Making of a Narco State," *Rolling Stone*, December 4, 2014.

85 Dasht-e Margo: John W. Whitney, *Geology, Water, and Wind in the Lower Helmand Basin, Southern Afghanistan*, Scientific Investigations Report no. 2006–5182 (Reston, Virginia: US Geological Survey, 2006), 5, 19.

89 the *emotion-recording apparatus*: Robert Graves, *Goodbye to All That* (London: Penguin Books, 1960), 240.

8

99 the most international destinations: Julien Lebel, "Turkish Airlines: An International Strategic Instrument for Turkey," *Études de l'Ifri*, IFRI (April 2020), 18.

99 *Angels of steel*: Michel Serres, *La légende des anges* (Paris: Flammarion, 1993), 8.

99 I started covering the war in Syria: Matthieu Aikins, "Whoever Saves a Life," *Matter*, September 15, 2014.

99 killing forty-five people: "6 Convicted for 2016 Istanbul Airport Attack That Killed 45," Associated Press, November 16, 2018.

99 authorities rely on automated profiling: Matthew Longo, *The Politics of*

Borders: Sovereignty, Security, and the Citizen after 9/11 (Cambridge, Cambridge University Press, 2017), 17, 141.

99　*The speed of some*: Tim Cresswell, *On the Move: Mobility in the Modern Western World* (London: Routledge, 2006), 240.

100　*a gift from God*: Patrick Kingsley, "Turkey Detains 6,000 over Coup Attempt as Erdoğan Vows to 'Clean State of Virus,'" *Guardian*, July 17, 2016.

100　fifty thousand people would be arrested: Howard Eissenstat, "Erdoğan as Autocrat: A Very Turkish Tragedy," Project on Middle East Democracy, April 12, 2017, 13; Human Rights Watch, "In Custody: Police Torture and Abductions in Turkey," October 12, 2017.

104　were building a massive border fence: Amnesty International, "The Human Cost of Fortress Europe: Human Rights Violations against Migrants and Refugees at Europe's Borders," July 9, 2014.

104　*seedy or out of sorts*: Isabel Burton and W. H. Wilkins, *The Romance of Isabel Lady Burton* (London: Hutchinson & Co., 1897), 542.

104　whose byproduct was the modern refugee: Peter Gatrell, *The Making of the Modern Refugee* (Oxford: Oxford University Press, 2013), 18; Michael Marrus, *The Unwanted* (New York: Oxford University Press, 1985), 15.

104　*the right to have rights*: Hannah Arendt, *The Origins of Totalitarianism* (New York: Harcourt, Brace & World, 1966), 296.

105　*It can be shown that this idea*: Immanuel Kant, "Perpetual Peace: A Philosophical Sketch," *Kant's Political Writings*, ed. Hans Reiss, trans. H. B. Nisbet (Cambridge: Cambridge University Press, 1971), 104.

105　*materially impossible*: Robert Schuman, "The Schuman Declaration— 9 May 1950."

105　*Over the past sixty years*: Herman Van Rompuy and José Manuel Durão Barroso, "From War to Peace: A European Tale," December 10, 2012.

105　*I would like to live*: Gérard de Nerval, *Voyage en Orient: Tome second* (Paris: Le Divan, 1927), 275.

105　Schengen and Dublin: Gregory Feldman, *The Migration Apparatus: Security, Labor, and Policymaking in the European Union* (Stanford: Stanford University Press, 2011), 61; David Scott FitzGerald, *Refuge beyond Reach: How Rich Democracies Repel Asylum Seekers* (Oxford: Oxford University Press, 2019), 170.

105 lost half its population: Dirk Hoerder, *Cultures in Contact: World Migrations in the Second Millennium* (Durham: Duke University Press, 2002), 339.

106 globalization's central tension: Sandro Mezzadra and Brett Neilson, *Border as Method, or, The Multiplication of Labor* (Durham: Duke University Press, 2013), 19.

106 *externalization*: Pinar Gedikkaya Bal, "The Effects of the Refugee Crisis on the EU-Turkey Relations," *European Scientific Journal* 12, no. 8 (March 2016); Stefan Alscher, "Knocking at the Doors of 'Fortress Europe': Migration and Border Control in Southern Spain and Eastern Poland," Center for Comparative Immigration Studies, working paper no. 126, November 2005.

106 the *harraga*, those who burn: Johanna Sellman, "A Global Postcolonial: Contemporary Arabic Literature of Migration to Europe," *Journal of Postcolonial Writing* 54, no. 6 (2018), 752; Hakim Abderrezak, "The Mediterranean Sieve, Spring and Seametery," in *Refugee Imaginaries: Research across the Humanities*, eds. Emma Cox et al. (Edinburgh: Edinburgh University Press, 2020), 373.

106 two continents whose vast gap: Branko Milanovic, *Capitalism, Alone: The Future of the System that Rules the World* (Cambridge: Belknap, 2019), 156.

107 *Tomorrow Europe*: BBC News, "Gaddafi Wants EU Cash to Stop African Migrants," August 31, 2010.

108 a playground for the communist elite: Kristen Ghodsee, *The Red Riviera: Gender, Tourism, and Postsocialism on the Black Sea* (Durham: Duke University Press, 2005), 85.

108 Such physical barriers had been rare: Reece Jones, *Violent Borders: Refugees and the Right to Move* (New York: Verso, 2016), 32.

108 the guards were assisted: Nadya Stoynova, Tihomir Bezlov, et al., "Human Smuggling in Bulgaria," in *Cross-Border Organised Crime: Bulgaria and Norway in the Context of the Migrant Crisis* (Sofia: Center for the Study of Democracy, 2017), 13; Michael Sontheimer and Barbard Supp, "Avenging East Germans Killed in Bulgaria," *Der Spiegel*, July 4, 2008.

108 *The theory then*: Rick Lyman, "Bulgaria Puts Up a New Wall, but This One Keeps People Out," *New York Times*, April 5, 2015.

108   there were only fifteen borders: Jones, *Violent Borders*, 88.

109   routinely beaten and robbed and forced back: Human Rights Watch, "Bulgaria: Pushbacks, Abuse at Borders," January 20, 2016.

109   *a near-mythical entity*: Kapka Kassabova, "Border Ghosts," *World Literature Today*, January 2018.

113   *Turn back from existence*: Rumi, *Masnavi*, book 2, lines 688, 700–701, 704, 706.

                                         9

122   Out of the more than twenty million: UNHCR, "Global Trends: Forced Displacement in 2015."

122   an instrument of soft power: Margaret Piper, Paul Power, and Graham Thom, "Refugee Resettlement: 2012 and Beyond," *New Issues in Refugee Research*, research paper no. 253, UNHCR (February 2013).

122   resettled one in twenty refugees: Gary Troeller, "UNHCR Resettlement: Evolution and Future Direction," *International Journal of Refugee Law* 14, no. 1 (January 2002).

122   the system was breaking down: Uğur Yıldız, *Tracing Asylum Journeys: Transnational Mobility of Non-European Refugees to Canada via Turkey* (London: Routledge, 2019), 44, 48.

123   thirty-five thousand Iraqis and Iranians: Kemal Kirişçi, "Turkey's New Draft Law on Asylum: What to Make of It?" in *Turkey, Migration and the EU*, eds. Seçil Paçacı Elitok and Thomas Straubhaar (Hamburg: Hamburg University Press, 2012) 66.

123   hunger strike at the asylum office: Noah Arjomand, "Afghan Exodus: Smuggling Networks, Migration and Settlement Patterns in Turkey," Afghanistan Analysts Network, September 10, 2016.

123   *In a refugee camp*: Dina Nayeri, *The Ungrateful Refugee* (Edinburgh: Canongate Books, 2019), 6, 8.

123   *It is not the voice*: Italo Calvino, *Invisible Cities*, trans. William Weaver (New York: Harcourt Brace, 1974), 135.

124   *refugee status determination*: UNHCR, "Refugee Status Determination: Identifying Who Is a Refugee," September 1, 2005, 112–13.

124   in Sweden: Swedish Migration Agency data for all positive decisions for Afghans in 2015 and 2017, excluding Dublin returns.

124   virtually guaranteed asylum: My analysis of Farah and Shireen's chances in Canada, the US, and Europe is based on published immigration statistics and interviews with asylum experts in those countries.

126   this quality of vulnerability: Mert Koçak, "Who Is 'Queerer' and Deserves Resettlement?: Queer Asylum Seekers and Their Deservingness of Refugee Status in Turkey," *Middle East Critique* (2020), 42.

126   *a new moral economy*: Didier Fassin, *Humanitarian Reason: A Moral History of the Present* (Berkeley: University of California Press, 2012), 87.

126   *wealthiest, luckiest, and strongest*: "Theresa May's Speech to the Conservative Party Conference—in Full," *The Independent*, October 6, 2015.

126   Western missionary groups were active: Shoshana Fine, *Borders and Mobility in Turkey: Governing Souls and States* (New York: Palgrave, 2018), 93, 120.

127   secret same-sex marriages: Ali Hamedani, "Gay Mullah Flees Iran over Secret Same-Sex Weddings," BBC News, June 7, 2016.

10

130   around three million: UNHCR, "Iran Factsheet," February 2016.

130   poem "Return": Author's translation of online text, excerpts, and discussion in Zuzanna Olszewska, *The Pearl of Dari: Poetry and Personhood among Young Afghans in Iran* (Bloomington: Indiana University Press, 2015), 57, 153.

130   Afghan refugees were increasingly unwelcome: Fariba Adelkhah and Zuzanna Olszewska, "The Iranian Afghans," *Iranian Studies* 40, no. 2 (2017), 154.

130   right to move and work freely was revoked: Human Rights Watch, "Unwelcome Guests: Iran's Violation of Afghan Refugee and Migrant Rights," November 2013.

130   their labor was still desired: Alessandro Monsutti, *War and Migration: Social Networks and Economic Strategies of the Hazaras of Afghanistan* (New York: Routledge, 2004), 127, 137.

131   forty-three refugee children: Olszewska, *The Pearl of Dari*, 22, 41–47.

132   traffic continued across: Adnan Çelik, "Challenging State Borders: Smuggling as Kurdish Infra-Politics During 'The Years of Silence,'" in *Kurds in Turkey: Ethnographies of Heterogeneous Experiences*, eds. Lucie Drechselová and Adnan Çelik (London: Lexington Books, 2019), 169, 177, 181.

132   supplying Europe through Bulgaria: Ryan Gingeras, *Heroin, Organized Crime, and the Making of Modern Turkey* (Oxford: Oxford University Press, 2014), 117, 234, 241.

135   relaunched its counterinsurgency campaign: International Crisis Group, "The PKK's Fateful Choice in Northern Syria," Middle East Report, no. 176, May 4, 2017.

11

138   a city of migrants: Barbara Pusch, "Bordering the EU: Istanbul as a Hotspot for Transnational Migration," in *Turkey, Migration and the EU: Potentials, Challenges and Opportunities*, eds. Seçil Paçacı Elitok and Thomas Straubhaar (Hamburg: Hamburg University Press, 2012), 174, 178.

138   neoliberal reforms: Cemal Burak Tansel, "National Neoliberalism in Turkey," *Dissent*, Summer 2019; Emily Canal, "The Cities with the Most Billionaires," *Forbes*, March 9, 2016.

138   *In the Ottoman period*: Ahmet İçduygu, "Turkey's Migration Transition and Its Implications for the Euro-Turkish Transnational Space," GTE Working Paper (Istituto Affari Internazionali), no. 7, April 2014.

138   the country had abolished: Kemal Kirişci "Border Management and EU-Turkish Relations: Convergence or Deadlock," CARIM Research Reports, Robert Schuman Centre for Advanced Studies, 2007.

138   the country's liberal visa policy: Ayşem Biriz Karaçay, "Shifting Human Smuggling Routes along Turkey's Borders," *Turkish Policy Quarterly* 15, no. 4 (Winter 2017), 97–108; Kemal Kirişci, "Will the Readmission Agreement Bring the EU and Turkey Together or Pull them Apart?," Centre for European Policy Studies, February 4, 2014.

138   *When our citizens*: Zeynep Özler, "Breaking the Vicious Circle in EU-

Turkey Relations: Visa Negotiations," *Turkish Policy Quarterly* 11, no. 1 (2012), 130.

139 *to cut the lines*: NATO, "Standing NATO Maritime Group Two Conducts Drills in the Aegean Sea," February 27, 2016.

146 much like the Chinese *fei'chien*: Patrick Radden Keefe, *The Snakehead: An Epic Tale of the Chinatown Underworld and the American Dream* (New York: Doubleday, 2009), 46.

146 migrant wages flow home: Peter Tinti and Tuesday Reitano, *Migrant, Refugee, Smuggler, Saviour* (Oxford: Oxford University Press, 2017), 68.

147 rural Afghanistan had survived: Alessandro Monsutti, *War and Migration: Social Networks and Economic Strategies of the Hazaras of Afghanistan* (New York: Routledge, 2004), 174, 185, 205.

12

150 *Very few of those escaped*: L'Abbé de Vertot, *The History of the Knights Hospitallers of St. John of Jerusalem, Volume 2* (New York: J.W. Leonard & Company, 1856), 104.

150 encouraged by Great Britain: Michael Llewellyn-Smith, *Ionian Vision: Greece in Asia Minor 1919–1922* (London: Hurst & Company, 1998), 29, 107, 288.

150 expelled one million refugees: Bruce Clark, *Twice a Stranger: The Mass Expulsions That Forged Modern Greece and Turkey* (Cambridge: Harvard University Press, 2006), 12, 18, 91.

150 *The condition of these people*: Henry Morgenthau, *I Was Sent to Athens* (New York: Doubleday, Doran & Company Inc., 1929), 101.

152 Izmir, the main staging point: Patrick Kingsley, *The New Odyssey: The Story of Europe's Refugee Crisis* (London: Guardian Books, 2016), 190.

154 a Canadian frigate: David Pugliese, "HMCS *Charlottetown* Returns Home from NATO Mission," *Ottawa Citizen*, January 13, 2017.

154 *He barely escaped*: James Stavridis, *Destroyer Captain: Lessons of a First Command* (Annapolis: Naval Institute Press, 2008), 128.

156 seventy-one migrants: Helene Bienvenu and Marc Santora, "They Let 71 People Die in a Stifling Truck. They Got 25 Years," *New York Times*, June 14, 2018.

160 more than thirty thousand migrants: Philippe Fargues, *Four Decades of Cross-Mediterranean Undocumented Migration to Europe: A Review of the Evidence*, International Organization for Migration, 2017.

160 *Nature sanitizes the killing floor*: Jason De León, *The Land of Open Graves: Living and Dying on the Migrant Trail* (Oakland: University of California Press, 2015), 83.

13

167 some forty thousand people landed: All figures on boat arrivals are taken from the UNHCR Operational Data Portal, Greece.

167 Greek authorities handed out pieces of paper: Katherina Rouzakou, "Nonrecording the 'European Refugee Crisis' in Greece: Navigating through Irregular Bureaucracy," *Focaal* 77 (March 2017).

168 unusual numbers of women and children: "Nationality of Arrivals to Greece, Italy and Spain: January–December 2015," Operational Data Portal, UNHCR, December 31, 2015; "Breakdown of Men–Women–Children among Sea Arrivals in Greece for the Period June 2015–January 2016," UNHCR, January 31, 2016; Tara Brian and Frank Laczko, "Fatal Journeys, Volume 2: Identification and Tracing of Dead and Missing Migrants," International Organization for Migration, 2016, 5.

168 *And the sea gave up*: Revelations 20:13, King James Bible.

168 *fed from the corpses*: George Tyrikos-Ergas, "Orange Life Jackets: Materiality and Narration in Lesvos, One Year after the Eruption of the 'Refugee Crisis,'" *Journal of Contemporary Archaeology* 3, no. 2 (2016), 230–31.

168 went viral on Facebook and Twitter: Francesco D'Orazio, "Journey of an Image: From a Beach in Bodrum to Twenty Million Screens Across the World," in *The Iconic Image on Social Media: A Rapid Research Response to the Death of Aylan Kurdi*, eds. Farida Vis and Olga Goriunova, Visual Social Media Lab, December 2015, 16; Simon Rogers, "What Can Search Data Tell Us About How the Story of Aylan Kurdi Spread around the World?," *The Iconic Image on Social Media*, 22.

168 His body ran on the continent's front pages: Lucía Abellán, "Una imagen que estremece la conciencia de Europa," *El País*, September 3, 2015;

Katie Hopkins, "Rescue Boats? I'd Use Gunships to Stop Migrants," *Sun*, April 17, 2015; Tom Wells, Nick Pisa, and Oliver Harvey, "SUN CAMPAIGN: Our Bid to Help Thousands of Kids Like Drowned Migrant Boy," *Sun*, September 3, 2015.

169 *the constant reports of car bombings*: Karl Ove Knausgaard, "Vanishing Point," *New Yorker*, November 17, 2015.

169 From a child, the miracle: Positive reception of refugees in Mauro Barisione, Asimina Michailidou, and Massimo Airoldi, "Understanding a Digital Movement of Opinion: The Case of #RefugeesWelcome," in *Information, Communication & Society* (December 2017), 5; Emma Graham-Harrison, Patrick Kingsley, Rosie Waites, and Tracy McVeigh, "Cheering German Crowds Greet Refugees after Long Trek from Budapest to Munich," *Guardian*, September 5, 2015; Dietrich Thränhardt, "Welcoming Citizens, Divided Government, Simplifying Media: Germany's Refugee Crisis, 2015–2017," in *Refugee News, Refugee Politics: Journalism, Public Opinion and Policymaking in Europe*, eds. Giovanna Dell'Orto and Irmgard Wetzstein (New York: Routledge, 2019), 15.

169 greatest movement of refugees by sea in history: The Mariel Boatlift in 1980 brought a hundred and fifty thousand in six months. After the Vietnam War, eight hundred thousand fled by sea, but over a period of two decades.

169 now a majority said: Tony Paterson, "Refugee Crisis: Germany's 'Welcome Culture' Fades as Thousands Continue to Arrive," *Independent*, October 7, 2015.

170 building razor-wire barriers: Ainhoa Ruiz Benedicto and Pere Brunet, "Building Walls: Fear and Securitization in the European Union," Centre Delàs d'Estudis per la Pau, September 2018.

170 On New Year's Eve in Cologne: quotes in Yermi Brenner and Katrin Ohlendorf, "Time for the Facts. What Do We Know about Cologne Four Months Later?," *Correspondent*, May 2, 2016.

170 *What would little Aylan*: Eugénie Bastié, "'Que serait devenu le petit Aylan s'il avait grandi?': *Charlie Hebdo* choque," *Le Figaro*, January 14, 2016.

170 *We must now secure*: "Schäuble will Autofahrer in Asylkrise zur Kasse bitten," *Die Welt*, January 16, 2016.

171 the administrative compound: For details see Antonis Vradis, Evie Papada, Joe Painter, and Anna Papoutsi, *New Borders: Hotspots and the European Migration Regime* (London: Pluto Press, 2019), 79–80; Silvan Pollozek and Jan Hendrik Passoth, "Infrastructuring European Migration and Border Control: The Logistics of Registration and Identification at Moria Hotspot," *Society and Space* 37, no. 4 (2019), 612.

173 EURODAC: Gregory Feldman, *The Migration Apparatus: Security, Labor, and Policymaking in the European Union* (Stanford: Stanford University Press, 2011), 134.

173 conditions were too atrocious: David Scott FitzGerald, *Refuge beyond Reach: How Rich Democracies Repel Asylum Seekers* (Oxford: Oxford University Press, 2019), 172.

174 database was strictly limited: Lehte Roots, "The New Eurodac Regulation: Fingerprints as a Source of Informal Discrimination," *Baltic Journal of European Studies*, vol. 5, no. 2 (2015), 109.

174 the Paris attackers: Anthony Faiola and Souad Mekhennet, "Tracing the Path of Four Terrorists Sent to Europe by the Islamic State," *Washington Post*, April 22, 2016.

174 merged with other databases: European Union Agency for Fundamental Rights, "Fundamental Rights and the Interoperability of EU Information Systems: Borders and Security," May 2017.

174 colonial officials: Chandak Sengoopta, *Imprint of the Raj: How Fingerprinting Was Born in Colonial India* (London: Macmillan, 2003), 51.

174 finally the metropole: Matthew Longo, *The Politics of Borders: Sovereignty, Security, and the Citizen after 9/11* (Cambridge, Cambridge University Press, 2017), 148.

174 an immense trove: Annie Jacobsen, *First Platoon: A Story of Modern War in the Age of Identity Dominance* (New York: Dutton, 2021), 294.

174 *Biopolitical tattooing*: Giorgio Agamben, "Non au tatouage biopolitique," *Le Monde*, January 10, 2004.

14

182 five thousand crammed: Emina Cerimovic, "Asylum Seekers' Hell in a Greek 'Hotspot,'" Human Rights Watch, November 30, 2017.

182   teetering on the edge of disaster: Bill Frelick, "Greece: Refugee 'Hotspots' Unsafe, Unsanitary," Human Rights Watch, May 19, 2016.

182   the fire spread swiftly: UNHCR "Fire at Reception Site on Lesvos Island, Greece," September 20, 2016.

183   *The queues have agency*: Behrouz Boochani, *No Friend but the Mountains: Writing from Manus Prison* (Toronto: House of Anansi, 2019), 197, 125.

184   EuroRelief: Background and decision to stay in Moria from interview with Stefanos Samiotakis, EuroRelief director, Mytilene, July 17, 2017.

184   *He has a plan*: Lori Golinghorst and Anita Schlabach, "Not Forgotten by God," https://lovinglesvos2016.wordpress.com/2016/03/01/not -forgotten-by-god/.

184   around eighty NGOs: Dina Siegel, *Dynamics of Solidarity: Consequences of the 'Refugee Crisis' on Lesbos* (The Hague: Eleven International Publishing, 2019), 75, 79, 85.

184   pulled out in protest: Euractiv, "MSF, Oxfam Pull Out of Lesbos Hotspot in Yet Another Blow to EU," March 24, 2016.

185   de facto control: European Center for Constitutional and Human Rights, "Case Report: EASO's Involvement in Greek Hotspots Exceeds the Agency's Competence and Disregards Fundamental Rights," April 2019.

185   The Greek government: Stathis Kouvelakis, "Borderland: Greece and the EU's Southern Question," *New Left Review*, no. 110 (March–April 2018), 7, 31.

185   *It's not mandatory*: Euractiv, "EU mulls plan to take charge of Europe's borders," December 5, 2015.

185   guards beside them: Apostolis Fotiadis, "New Security on Greek Islands Reduces Access," *New Humanitarian*, June 15, 2016.

186   Even Pope Francis: Jim Yardley, "Pope Francis Takes 12 Refugees Back to Vatican after Trip to Greece," *New York Times*, April 16, 2016.

15

188   once the headquarters: Bruce Clark, *Twice a Stranger: The Mass Expulsions That Forged Modern Greece and Turkey* (Cambridge: Harvard University Press, 2006), 28.

188 *someone will remember us*: Sappho, *If Not, Winter: Fragments of Sappho*, trans. Anne Carson (New York: Knopf, 2002), 298.

189 thousands of refugees had been camped: Nicolas Niarchos, "An Island of Refugees," *New Yorker*, September 16, 2015.

190 crowded with sick: Esther Lovejoy, *Certain Samaritans* (New York: Macmillan, 1927), 174.

190 villages were still referred: Leonidas Oikonomakis, "Solidarity in Transition: The Case of Greece," *Solidarity Mobilizations in the "Refugee Crisis,"* ed. Donatella della Porta (London: Palgrave Macmillan, 2018), 77.

191 *unmix the populations of the Near East*: Clark, *Twice a Stranger*, 93.

195 *The island is a prison*: Christos Ikonomou, *Good Will Come from the Sea*, trans. Karen Emmerich (Brooklyn: Archipelago, 2019), 40.

195 *We welcomed refugees*: Joanna Kakissis, "For These Greek Grandmas, Helping Migrants Brings Back Their Own Past," National Public Radio, April 15, 2016.

195 *If we continue*: "Barcelona, París, Lesbos y Lampedusa piden a los Estados que 'no den la espalda' a las ciudades refugio," *Público*, September 13, 2015.

196 *You know, after the prizes*: Interview with Giorgos Tyrikos-Ergas, Kalloni, July 13, 2017.

196 the betting shops: Kate Brady, "Nobel Peace Prize: Who Will Win?," *Deutsche Welle*, October 7, 2016.

196 *felt pity at first*: John Steinbeck, *The Grapes of Wrath* (New York: Penguin Books, 2002), 434.

196 Tourists hardly visited: Stanislav Ivanov and Theodoros A. Stavrinoudis, "Impacts of the Refugee Crisis on the Hotel Industry: Evidence from Four Greek Islands," *Tourism Management* 67 (August 2018), 214–23.

196 the Chios camp was firebombed: Helena Smith and Patrick Kingsley, "Far-Right Group Attacks Refugee Camp on Greek Island of Chios," *Guardian*, November 18, 2016.

196 parents chained shut: Mary Harris, "Lesbos Parents Lock Schools to Keep Refugees Out," *Greek Reporter*, October 9, 2016.

199 *His power was*: Edward W. Said, *Orientalism* (New York: Vintage, 1994), 160.

16

201   *It is a kind of duty*: George Orwell, *The Road to Wigan Pier* (New York: Harcourt, 1958), 17, 33.

202   *Effective border protection*: Tony Abbott, "Address to the Alliance of European Conservatives and Reformists, Lobkowicz Palace, Prague, Czech Republic," September 17, 2016.

202   dengue and Zika viruses: James Cavallaro, Diala Shamas, Beth Van Schaack, et al., "Communiqué to the Office of the Prosecutor of the International Criminal Court Under Article 15 of the Rome Statute," Stanford Human Rights Center, February 14, 2017, 52.

202   the policy had been effective: "Polls Apart: How Australian Views Have Changed on 'Boat People,'" Lowy Institute, February 19, 2019; Janet Phillips, "Boat Arrivals and Boat 'Turnbacks' in Australia Since 1976: A Quick Guide to the Statistics," Research Paper Series 2016–17, Parliament of Australia, January 17, 2017.

203   they were unsafe after dark: Human Rights Watch, "Greece: Dire Risks for Women Asylum Seekers," December 15, 2017.

203   scabies and varicella: Maaike Hermans et al., "Healthcare and Disease Burden Among Refugees in Long-Stay Refugee Camps at Lesbos, Greece," *European Journal of Epidemiology* 32 (2017), 853.

203   *Mass international migration*: Paul Collier, *Exodus: How Migration Is Changing Our World* (Oxford: Oxford University Press, 2013), 271, 26.

204   converging with the West: Branko Milanovic, *Global Inequality: A New Approach for the Age of Globalization* (Cambridge: Harvard University Press, 2018), 166.

204   Looking at the problem in an absolute sense: David Woodward, "Incrementum ad Absurdum: Global Growth, Inequality and Poverty Eradication in a Carbon-Constrained World," *World Social and Economic Review*, no. 4 (2015); Jason Hickel, "The Contradiction of the Sustainable Development Goals: Growth versus Ecology on a Finite Planet," *Sustainable Development* 27 (2019).

204   *When we show up*: Mark Frankel, "From Morningside Heights to the Maldives," Columbia Law School, May 12, 2016.

206   suicide attempts: Barak Kalir and Katerina Rozakou, "'Giving Form to

Chaos': The Futility of EU Border Management at Moria Hotspot In Lesvos," *Society and Space*, November 16, 2016.

209 *From this exile*: Farhad Darya, "Salaam Afghanistan."

210 *That is understandable*: Reuters, "Germany Offers Afghanistan Help to Take Back Migrants," February 1, 2016.

210 the Joint Way Forward: Marissa Quie and Hameed Hakimi, "The EU and the Politics of Migration Management in Afghanistan," Chatham House, November 2020, 4.

211 *no sympathy*: Yalda Hakim, "President Ghani Calls for Afghans to Remain in Country," BBC News, March 31, 2016.

213 *There are no refugees*: Eliza Mackintosh, "Conditions Worsen for Europe's Refugees as Temperatures Plummet," CNN, January 13, 2017.

213 asphyxiated by fumes: Eva Cossé, "Death and Despair in Lesbos," Human Rights Watch, February 3, 2017.

213 *If one investigated*: Behrouz Boochani, *No Friend but the Mountains: Writing from Manus Prison* (Toronto: House of Anansi, 2019), 212, 214, 318.

214 *mental health emergency*: Médecins sans Frontières, "Confronting the Mental Health Emergency on Samos and Lesvos," October 2017, 7.

214 *Why don't they just kill us?*: Infomobile, "Moria/Lesbos: Tear Gas and Beatings Continue while Families Wait in the Mud All the Night," October 7, 2015; Joanna Slater, "Don't Send Anyone Else. This Is a Very Difficult Road," *Globe and Mail*, September 4, 2015.

17

225 *Before you came*: Faiz Ahmed Faiz, *The True Subject: Selected Poems of Faiz Ahmed Faiz*, trans. Naomi Lazard (Princeton: Princeton University Press, 1988), 33.

230 a riot against immigrants: Daniel Trilling, *Lights in the Distance* (London: Picador, 2018), 124, 181.

231 The city exploded in recurring protests: Costis Hadjimichalis, "From Streets and Squares to Radical Political Emancipation? Resistance Lessons from Athens during the Crisis," *Human Geography* 6, no. 2 (July 2013), 130.

231 The anarchists lived here: Nicholas Apoifis, "'FUCK MAY 68, FIGHT

NOW!' Athenian Anarchists & Anti-Authoritarians: Militant Ethnography & Collective Identity Formation," (MA thesis, Macquarie University, 2014), 14, 107.

231 Diktyo, a network: Natasha King, *No Borders: The Politics of Immigration Control and Resistance* (London: Zed Books, 2016), 58–60.

232 the Greek solidarity movement: Katerina Rozakou, "Socialities of Solidarity: Revisiting the Gift Taboo in Times of Crises," *Social Anthropology* 24, no. 2 (2016), 185; Giorgos Maniatis, "From a Crisis of Management to Humanitarian Crisis Management," *South Atlantic Quarterly* 117, no. 3 (October 2018), 908.

232 occupied an empty hotel: Olga Lafazani, "Homeplace Plaza: Challenging the Border between Host and Hosted," *South Atlantic Quarterly* 117, no. 3 (October 2018), 896.

233 broke Syriza's resistance: Yanis Varoufakis, *Adults in the Room: My Battle With Europe's Deep Establishment* (New York: Farrar, Straus and Giroux, 2017), 457.

233 promised to end: Preethi Nallu, "Greece Outlines Radical Immigration Reforms," Al Jazeera, March 5, 2015.

234 fifty thousand refugees: Amnesty International, "Trapped in Greece: An Avoidable Refugee Crisis," April 18, 2016.

234 around two thousand migrants: Valeria Raimondi, "For 'Common Struggles of Migrants and Locals': Migrant Activism and Squatting in Athens," *Citizenship Studies* 23, no. 6 (2019), 566–67.

234 So far, the government: interview with Aliki Papachela, Athens, June 21, 2017.

234 To read the Greek papers: Lina Giannarou, "Ta Gkriza Shmeia Ths Katalhyhs Toy City Plaza," *Kathimerini*, October 27, 2016.

238 *a virtual paradise*: Aryn Baker, "Greek Anarchists Are Finding Space for Refugees in Abandoned Hotels," *Time*, November 3, 2016.

240 income by 40 percent: Eirini Andriopoulou, Alexandros Karakitsios, and Panos Tsakloglou, "Inequality and Poverty in Greece: Changes in Times of Crisis," Hellenic Observatory, GreeSE paper no. 116, 2016.

240 besieged by drug gangs: Giorgos Poulimenakos and Dimitris Dalakoglou, "Hetero-Utopias: Squatting and Spatial Materialities of Resistance in Athens at Times of Crisis," in *Critical Times in Greece:*

Anthropological Engagements with the Crisis, eds. Dimitris Dalakoglou and Giorgos Agelopoulos (New York: Routledge, 2018), 174, 179.

241   *We take the responsibility*: "Responsibility Claim for the Execution of Mafioso Habibi in June, Exarhia Area—Athens, Greece," July 14, 2016, https://actforfree.nostate.net/?p=24330.

241   *Mi vida va prohibida*: Manu Chao, "Clandestino."

242   *crimes of solidarity*: Liz Fekete, Frances Webber, Anya Edmond-Pettit, "Humanitarianism: The Unacceptable Face of Solidarity," Institute of Race Relations, 2017.

242   a *French Underground Railroad*: Adam Nossiter, "A French Underground Railroad, Moving African Migrants," *New York Times*, October 4, 2016.

242   an activist from Barcelona: "Greece: Suspended Sentence for Spanish Activist Is 'Decisive' for Decriminalising Solidarity with Migrants and Refugees," Statewatch, December 19, 2018.

18

245   a boy buried with his tortoiseshell lyre: Stelios Psaroudakes, "A Lyre from the Cemetery of the Acharnian Gate, Athens," in *Music Archeology in Contexts*, eds. Ellen Hickman, Arnd Both, and Ricardo Eichmann (Rahden: Verlag Marie Leiderdorf, 2006), 59.

245   a terra-cotta cistern: Luis E. Navia, *Diogenes of Sinope: The Man in the Tub* (Westport: Greenwood Press, 1998), 19, 22, 123.

245   see Athens decline: Population figures from Wilfried Nippel, *Ancient and Modern Democracy: Two Concepts of Liberty?*, trans. Keith Tribe (Cambridge: Cambridge University Press, 2015), 10; Kevin Andrews, *Athens Alive, or The Practical Tourist's Companion to the Fall of Man* (Athens: Hermes Publications, 1979), 198.

245   Kotzia Square: Gabriella Arrigoni et al, "Online Visual Dialogues about Place: Using the Geostream Tools to Identify Heritage Practices on Photo-sharing Social Media," CoHERE, March 28, 2017, 9, 26.

247   informal red-light district: Andriopoulos Themis, "Brothels: Houses Which Stood the Test of Time," *Athens Social Atlas*, eds. Thomas Maloutas and Stavros Spyrellis, May 2015.

247   used by male prostitutes: Will Horner, "Afghan Asylum Seekers Resort to Sex Work in Athens," Al Jazeera, January 16, 2017.

247   Afghanistan stood out: Phillip Connor and Jens Manuel Krogstad, "Europe Sees Rise in Unaccompanied Minors Seeking Asylum, with Almost Half from Afghanistan," Pew Research Center, May 10, 2016.

247   Some countries, like Sweden: Martin Nyman and Peter Varga, "Country Report: Sweden, 2019 Update," Asylum Information Database, 2019, 88.

248   kids were being held in jails: Human Rights Watch, "'Why Are You Keeping Me Here?': Unaccompanied Children Detained in Greece," September 2016, 9.

249   *The whole Jungle seems empty*: Song by Abdullah, translated from Pashto by Thomas Wide and recorded by Celeste Cantor-Stephens, "New Jungle Music: A Fieldwork Study of the Migrant Jungle of Calais," (MSt Music, Oxford, 2015), 51.

249   *long summer of migration*: Sandro Mezzadra, "In the Wake of the Greek Spring and the Summer of Migration," *South Atlantic Quarterly* 117, no. 3 (October 2018), 927.

250   faced harsh violence: Human Rights Watch, "Hungary: Migrants Abused at the Border," July 13, 2016.

19

252   admiral of the Ionian Fleet: Interview with Greek naval officer, Athens, 2017.

252   a shantytown back then: Olga Lafazani, "A Border within a Border: The Migrants' Squatter Settlement in Patras as a Heterotopia," *Journal of Borderlands Studies* 28, no. 1 (2013), 2.

262   a war broke out: Behzad Yaghmaian, *Embracing the Infidel: Stories of Muslim Migrants on the Journey West* (New York: Delta, 2006), 221–25; Matthew Carr, *Fortress Europe: Dispatches from a Gated Continent* (London: Hurst, 2015), 102–3.

265   *catastrophe which keeps piling*: Walter Benjamin, "Theses on the Philosophy of History," in *Illuminations*, trans. Harry Zohn (New York: Schocken Books, 2007), 257.

267 *If the hand*: Shahram Khosravi, *"Illegal" Traveller: An Auto-Ethnography of Borders* (New York, Palgrave Macmillan, 2010), 60, 62.

20

272 *mobile, permeable, and discontinuous*: Sandro Mezzadra and Brett Neilson, *Border as Method, or, the Multiplication of Labor* (Durham: Duke University Press, 2013), 269.

273 That meant the world system: Charles Heller, Lorenzo Pezzani, and Maurice Stierl, "Toward a Politics of Freedom of Movement," *Open Borders*, ed. Reece Jones (Athens: University of Georgia Press, 2019), 59.

273 more than five thousand: *Fatal Journeys: Volume 3, Part 1*, eds. Frank Laczko, Ann Singleton, and Julia Black, IOM, September 2017, 6.

274 *there's plenty of money*: Kostas Kostis, *History's Spoiled Children: The Formation of the Modern Greek State* (London: Hurst, 2018), 399.

274 even further into debt: Truth Committee on Public Debt, "Preliminary Report," Hellenic Parliament, June 18, 2015, 20; IMF, "Greece: 2017 Article IV Consultation," February 7, 2017, 6.

276 the event was moved: Mary Harris, "U.S. President Obama's Speech Cancelled at Pnyx Hill," *Greek Reporter*, November 11, 2016.

277 *I want to commend*: Barack Obama, "Remarks by President Obama at Stavros Niarchos Foundation Cultural Center in Athens, Greece," November 16, 2016.

278 *The verdict of the Greek people*: Griff Witte, "Greeks Emphatically Reject Austerity, Elect Syriza in Historic Vote," *Washington Post*, January 25, 2015.

279 last days of the autonomous zone: Molly Crabapple, "The Attack on Exarchia, an Anarchist Refuge in Athens," *New Yorker*, January 20, 2020.

279 *a revival of a new generation*: "Mitsotakis Slams 'Inconceivable Tolerance' of Gov't for Anarchist Violence," *Ekathimerini*, January 22, 2018.

279 second-most popular neighborhood: Nikos Roussanoglou, "Plaka and Exarchia Top Airbnb Chart," *Kathimerini*, October 26, 2017.

279 *So what if they douse*: Faiz Ahmed Faiz, "Prison Nightfall," trans. Ted Genoways, "'Let Them Snuff Out the Moon': Faiz Ahmed Faiz's Prison Lyrics in Dast-e Saba," *Annual of Urdu Studies* (2004), 115.

280 Kierkegaard: Søren Kierkegaard, *Works of Love*, trans. Howard and Edna Hong (New York: Harper Perennial, 2009), 344.

21

282 *Near, far, wherever you are*: Céline Dion, "My Heart Will Go On."

285 *story-telling animal*: Alasdair MacIntyre, *After Virtue: A Study in Moral Theory* (University of Notre Dame, 1981), 216.

286 *Whatever longing subtracts*: Rumi, *Masnavi*, book 5, line 3683.

287 richer diets: Anna Lagia, "Health Inequalities in the Classical City," *L'apport de la paléoanthropologie funéraire aux sciences historiques*, ed. Anne-Catherine Gillis (Lille: Septentrion, 2014), 103.

287 Many of the people here must have been slaves: Louise Cilliers, "Burial Customs, the Afterlife and the Pollution of Death in Ancient Greece," *Acta Theologica* 26, no. 2 (March 2010) 55; Paul Cartledge, *Democracy: A Life* (Oxford: Oxford University Press, 2016), 138.

287 *For that some should rule*: Aristotle, *The Politics of Aristotle*, trans. Benjamin Jowett (Oxford: Clarendon, 1885), 7.

287 *The effect of the necessity*: Bernard Williams, *Shame and Necessity* (Berkeley: University of California Press, 1993), 124.

290 Gloria Anzaldúa wrote: *Borderlands/La Frontera: The New Mestiza* (San Francisco: Aunt Lute Books, 1987), 30, 85.

291 *See the signs*: "Hum Yaar Hain Tumhare," from the film *Haan Maine Bhi Pyaar Kiya*.

About the Author

Matthieu Aikins has reported from Afghanistan and the Middle East since 2008. He is a contributing writer for the *New York Times Magazine* and a contributing editor at *Rolling Stone*, and has won numerous honors, including the George Polk and Livingston awards. He has been a fellow at Type Media Center, New America, the Council on Foreign Relations, and the American Academy in Berlin. Matthieu grew up in Nova Scotia, and has a master's degree in Near Eastern studies from New York University. *The Naked Don't Fear the Water* is his first book.